PROMPT ENGINEERING

The Author

Dr. S. Anandamurugan obtained his Bachelor's degree in Electrical and Electronics Engineering from Bharathiyar University and Masters's Degree in Computer Science and Engineering from Madurai Kamaraj University. He completed a Ph.D. in Wireless Sensor Networks at Anna University, Chennai. He has 24 years of teaching experience. He is currently working as an Associate Professor and Head of the Information Technology department of Kongu Engineering College. He got 10 awards. He has written 102 books under various titles and publications. Four of his books are prescribed as Text/Reference books by many leading Universities and Autonomous Colleges in India and Abroad.

PROMPT ENGINEERING

Dr. S. Anandamurugan
Associate Professor and Head
Department of Information Technology,
Kongu Engineering College, Perundurai, Erode-638060

2026

Associated Publishing Company
A Division of
Astral International Pvt. Ltd.
New Delhi – 110 002

ISBN: 978-93-5919-352-6

Published by : **Associated Publishing Company**
A Division of
Astral International Pvt. Ltd.
– ISO 9001:2015 Certified Company –
4736/23, Ansari Road, Darya Ganj
New Delhi-110 002
Ph. 011-43549197, 8130496929
E-mail: info@astralint.com
Website: www.astralint.com

Printed at : **Replika Press Pvt. Ltd.**

PREFACE

Prompt engineering is the art and science of designing and optimizing prompts to guide AI models, particularly LLMs, towards generating the desired responses. By carefully crafting prompts, you provide the model with context, instructions, and examples that help it understand your intent and respond in a meaningful way.

Prompt engineering is carried out by prompt engineers to leverage the natural language processing capabilities of the AI model to generate better results. Organizations are typically looking to achieve the following objectives with prompt engineering techniques:

- Improved quality control over AI-generated results
- Mitigate any biases in the output from the AI model
- Generate personalized content for very specific domains
- Get consistent results that are relevant to the expectations of the user.

Features:

- Student friendly –written in a clear, concise and lucid manner
- A sincere attempt has been made to introduce the basic concepts
- Each chapter is organized into small sections that address key topics
- In-depth coverage and good style are emphasized
- Motivates the unmotivated
- Covers both elementary as well as advanced concepts
- Logical flow of concepts starting from the preliminary topics to the major topics

Dr. S. Anandamurugan

CONTENTS

CHAPTER 1

INTRODUCTION TO PROMPT ENGINEERING

Artificial Intelligence models, particularly large language models, are designed to generate human-like text based on the input they receive. However, the quality, accuracy, and usefulness of their responses depend heavily on how they are prompted. Prompt engineering is the practice of crafting effective input prompts to guide an AI model toward producing desired and meaningful outputs.

This section explores what prompt engineering is, why it is important, and how it is applied in artificial intelligence and large language models.

What is Prompt Engineering?

Introduction to Prompt Engineering

Prompt engineering is the process of carefully designing and structuring input prompts to effectively interact with artificial intelligence (AI) models, particularly Large Language Models (LLMs) such as GPT-4, Claude, Gemini, and others. It involves formulating instructions, queries, or context in a way that guides the AI to generate accurate, relevant, and high-quality responses.

As AI models rely on text-based input to understand and generate responses, the way a prompt is written significantly affects the output. Well-structured prompts lead to more precise, coherent, and useful answers, while poorly designed prompts can result in vague, incorrect, or misleading responses.

Prompt engineering is an essential skill for developers, researchers, businesses, educators, content creators, and AI enthusiasts who seek to harness the full potential of AI for various applications.

How Does Prompt Engineering Work?

Prompt engineering works by leveraging the language model's training data, linguistic patterns, and reasoning abilities to generate responses that align with user expectations. AI models function by predicting the most likely next word based on context, and well-crafted prompts help direct the AI's focus toward the desired response.

The key elements that influence AI responses include:

1. Prompt Clarity: Clearly defining what you want the AI to do improves response accuracy.
2. Context and Background Information: Providing relevant details enhances the model's understanding.
3. Instructional Precision: Specific instructions yield more controlled and structured outputs.
4. Prompt Formatting and Structure: Using bullet points, examples, and structured prompts improves readability.
5. Iteration and Refinement: Modifying prompts based on AI responses helps optimize outcomes.

For example, a vague prompt like "Tell me about climate change" may result in a broad or generic response. However, a refined prompt like "Explain the causes and effects of climate change in simple terms for a high school student" ensures a more targeted and useful answer.

Why is Prompt Engineering Important?

Prompt engineering is crucial because it maximizes the effectiveness of AI systems by improving response quality and ensuring that AI-generated content aligns with specific objectives. Below are some key reasons why prompt engineering is important:

1. Enhances AI Response Quality

Well-structured prompts help AI models produce more accurate, coherent, and contextually appropriate responses. This is essential in areas such as business automation, research, education, and creative writing.

2. Reduces AI Mistakes and Misinterpretations

AI models may misinterpret vague or ambiguous prompts, leading to incorrect or misleading information. Proper prompt engineering helps minimize these issues by providing clear guidance to the AI.

3. Saves Time and Improves Efficiency

By crafting precise prompts, users can obtain high-quality results faster without needing multiple revisions or re-iterations. This is particularly useful for tasks such as content creation, programming, data analysis, and chatbot interactions.

4. Optimizes AI for Specific Use Cases

Different industries and domains have unique requirements. Prompt engineering helps customize AI responses for specific fields, such as:

- Marketing: Generating engaging ad copies and product descriptions.
- Education: Creating interactive learning experiences.
- Healthcare: Assisting with medical documentation and research.
- Programming: Writing and debugging code efficiently.

5. Enables Complex AI Interactions

Advanced prompting techniques such as Chain-of-Thought (CoT) prompting, few-shot learning, and role-based prompting allow AI to handle complex tasks, reasoning-based queries, and contextual understanding.

6. Helps Control Bias and Ethical Concerns

By designing prompts carefully, users can mitigate bias, ensure fair AI responses, and avoid generating misinformation. Ethical prompt engineering is crucial for maintaining responsible AI usage.

Types of Prompting in Prompt Engineering

1. Instruction-Based Prompting

This involves giving AI explicit commands or instructions to achieve a desired outcome.

- Example: *"Summarize this article in three sentences."*

2. Role-Based Prompting

The AI is assigned a specific role or persona to tailor responses accordingly.

- Example: *"You are a legal expert. Explain the basics of contract law to a business owner."*

3. Few-Shot Prompting

A few examples are provided in the prompt to guide AI in understanding patterns.

- Example:
  - » Input: "John loves apples. Mary loves oranges."
  - » Few-shot prompt: "John enjoys apples. Mary enjoys ____?"
  - » AI Output: "oranges"

4. Chain-of-Thought Prompting

Encouraging AI to break down complex problems by explaining its thought process step by step.

- Example: *"Solve this math problem step by step: If a car travels 60 miles in 2 hours, what is its speed?"*

5. Zero-Shot Prompting

The AI is expected to perform a task without any prior examples, relying solely on its learned knowledge.

- Example: *"Translate this sentence into Spanish: 'How are you?'"*

Real-World Applications of Prompt Engineering

Prompt engineering has a wide range of applications across different industries. Some key areas where it is used include:

1. Business and Marketing

- Creating compelling advertisements and product descriptions.
- Automating customer support chatbots with improved response accuracy.
- Generating email marketing content and personalized customer messages.

2. Education and Research

- Assisting students with homework, tutoring, and academic writing.
- Summarizing research papers and generating study guides.
- Creating interactive educational quizzes and assessments.

3. Software Development and Code Generation

- Writing, debugging, and optimizing code snippets in various programming languages.
- Assisting with API documentation and automated script generation.
- Enhancing software development workflows with AI-assisted tools.

4. Content Creation and Journalism

- Assisting in article writing, news summaries, and storytelling.
- Generating blog posts, creative fiction, and social media content.
- Helping authors with idea brainstorming and manuscript refinement.

5. Healthcare and Medical Assistance

- Generating medical summaries and patient documentation.
- Assisting with diagnostic guidance and research (not a substitute for professional advice).
- Automating health-related chatbots for answering common medical queries.

6. Legal and Compliance

- Summarizing legal documents and contracts.
- Assisting in policy and compliance documentation.

- Providing basic legal explanations (without replacing professional legal advice).

Future of Prompt Engineering

As AI technology advances, prompt engineering will continue to evolve to support more sophisticated and nuanced interactions. Some key future developments include:

1. Automated Prompt Optimization – AI systems will be able to refine and adjust prompts dynamically for better performance.
2. Multi-Modal Prompting – AI models will process text, images, audio, and video inputs together, requiring more advanced prompting techniques.
3. Personalized AI Assistants – Prompt engineering will enhance context-aware AI assistants for individuals and businesses.
4. Stronger Ethical and Bias Controls – AI responses will be fine-tuned to minimize misinformation and enhance fairness.
5. Integration with No-Code and API-Based Platforms – Prompt engineering will be embedded in low-code and no-code tools, making AI more accessible.

Conclusion

Prompt engineering is a critical skill for optimizing AI interactions, ensuring accuracy, and improving response quality. By understanding how AI models process prompts, users can craft more effective, structured, and goal-oriented inputs for different applications.

As AI continues to shape various industries, mastering prompt engineering will be essential for businesses, educators, developers, and researchers looking to harness AI's full potential.

Importance of Prompt Engineering

Introduction

Prompt engineering plays a critical role in how effectively Artificial Intelligence (AI) models, particularly Large Language Models (LLMs), process and generate responses. As AI technology becomes increasingly integrated into various industries, the ability to craft precise and optimized prompts is essential for obtaining high-quality, relevant, and accurate outputs.

AI models, including GPT-4, Claude, Gemini, and others, rely on prompts as their primary means of understanding user intent. A well-structured prompt guides the AI in producing useful, contextually appropriate, and structured responses, while a poorly designed prompt may result in vague, inaccurate, or misleading outputs.

Why is Prompt Engineering Important?

The importance of prompt engineering stems from its ability to:

1. Improve the quality of AI-generated responses.
2. Reduce errors, ambiguity, and inconsistencies in AI outputs.
3. Optimize AI models for specific applications across different industries.
4. Increase efficiency and save time by reducing trial and error.
5. Control biases and ensure ethical AI usage.
6. Enhance AI's ability to handle complex reasoning and problem-solving tasks.

Let us explore each of these aspects in detail.

1. Improves AI Response Quality

AI models generate responses based on statistical patterns and probabilities rather than true comprehension. Without proper guidance, an AI may produce irrelevant, generic, or incomplete answers.

Well-structured prompts help guide the AI's response by clearly specifying the task, context, and desired output. This ensures that the generated content is:

- Accurate: The AI provides factual and correct information.
- Contextually Relevant: The response aligns with the user's intent.
- Well-Formatted: The output follows a logical structure for better readability.
- Creative and Detailed: The AI produces more insightful and original answers.

Example of Poor vs. Well-Engineered Prompt

Poor Prompt: *"Tell me about marketing."*

- The response may be too broad or general.

Well-Engineered Prompt: *"Explain digital marketing strategies used by small businesses to increase customer engagement, and provide three real-world examples."*

- The AI now understands the specific topic, expected depth of information, and need for examples, leading to a better-quality response.

2. Reduces Errors and Ambiguity in AI Outputs

AI models lack common sense and do not inherently understand ambiguity. A vague or unclear prompt can result in:

- Unreliable or inconsistent answers.
- Incomplete information that does not address the user's needs.
- Responses that misinterpret the intended question.

Prompt engineering helps eliminate these issues by ensuring that the AI receives clear and specific instructions.

Example of an Ambiguous vs. Clear Prompt

Ambiguous Prompt: *"Explain Python."*

The AI may not know if the user is referring to the programming language or the snake species.

Clear Prompt: *"Provide an introduction to Python programming, explaining its key features, advantages, and common applications."*

- The AI now understands the request unambiguously, leading to a relevant and precise response.

3. Optimizes AI for Specific Applications and Industries

Different industries have unique requirements, and AI needs to be adapted accordingly. Prompt engineering customizes AI interactions for various sectors, such as:

- Business and Marketing: Creating high-quality advertisements, customer engagement strategies, and market analysis.
- Education: Assisting with lesson planning, generating quizzes, and simplifying complex topics for students.
- Healthcare: Providing patient education, summarizing medical literature, and assisting in documentation.
- Programming: Writing and debugging code, explaining algorithms, and automating software development tasks.
- Creative Writing: Generating compelling stories, poetry, and content for books and blogs.

By designing prompts tailored to industry-specific needs, businesses and professionals can maximize the benefits of AI in their respective domains.

Example: AI in Healthcare vs. AI in Marketing

Healthcare Prompt: *"Summarize the latest research on Alzheimer's disease treatment and explain it in simple terms for non-medical readers."*

Marketing Prompt: *"Generate a social media marketing strategy for a new fitness brand targeting millennials, including key messaging, ad formats, and engagement tactics."*

Both prompts are designed to optimize AI responses for specific fields, ensuring better results for different users.

4. Saves Time and Enhances Efficiency

Without prompt engineering, users may spend significant time refining and rephrasing queries to obtain the desired AI response. By crafting effective prompts from the start, users can:

- Reduce the number of interactions needed to get the correct answer.
- Minimize back-and-forth corrections in AI-assisted workflows.
- Increase productivity by obtaining well-structured responses quickly.

This is particularly useful in business automation, coding, research, and content generation, where efficiency is crucial.

Example: Time-Saving in Content Generation

Poor Prompt: *"Write about climate change."*

- The AI might generate a general and unfocused response, requiring multiple revisions.

Optimized Prompt: *"Write a 500-word article on the impact of climate change on coastal cities, including three real-world examples and potential solutions."*

- The AI now produces a high-quality article on the first attempt, saving time and effort.

5. Controls Bias and Ensures Ethical AI Usage

AI models learn from vast datasets, which may contain biased or misleading information. Without careful prompt engineering, AI can produce:

- Unintentionally biased or discriminatory responses.
- Factually incorrect or misleading information.
- Harmful or offensive outputs.

By designing prompts responsibly, users can help ensure that AI provides fair, accurate, and ethical responses.

Example: Mitigating Bias in AI Responses

Potentially Biased Prompt: *"Why are electric cars better than gasoline cars?"*

- This assumes electric cars are superior without considering counterarguments.

Neutral and Ethical Prompt: *"Compare the advantages and disadvantages of electric cars and gasoline cars, considering factors like cost, environmental impact, and performance."*

- This ensures a balanced and objective response, reducing potential bias.

6. Enables AI to Handle Complex Reasoning and Problem-Solving

Advanced prompting techniques like Chain-of-Thought (CoT) prompting, few-shot learning, and self-consistency prompting allow AI to solve complex problems step by step.

For example, AI can:

- Break down mathematical problems into logical steps.

- Explain legal concepts with reasoning.
- Generate multi-step business strategies.

Example: Step-by-Step Reasoning with Chain-of-Thought Prompting

Simple Prompt: *"Solve 24 divided by 2 plus 6 times 3."*

- AI might provide a direct numerical answer without explanation.

Optimized Prompt: *"Solve 24 divided by 2 plus 6 times 3 step by step, explaining each calculation clearly."*

- AI breaks down the problem logically, improving transparency and learning.

Conclusion

Prompt engineering is a crucial skill for enhancing AI interactions, improving response accuracy, and ensuring ethical AI usage. It plays a vital role in business, education, healthcare, programming, content creation, and beyond.

By mastering prompt engineering, users can:

1. Maximize AI efficiency and productivity.
2. Reduce misinformation and bias.
3. Optimize AI for industry-specific applications.
4. Enhance AI's ability to solve complex problems.

As AI continues to evolve, effective prompt engineering will remain essential for unlocking the full potential of artificial intelligence across various domains.

Applications of Prompt Engineering in AI and Large Language Models

Introduction

Prompt engineering is a crucial technique that enhances the capabilities of Artificial Intelligence (AI) models, particularly Large Language Models (LLMs) such as GPT-4, Gemini, Claude, and LLaMA. These models rely on natural language inputs (prompts) to generate text-based outputs in response to user queries. The quality and specificity of a prompt directly influence the accuracy, relevance, and effectiveness of the AI's response.

Prompt engineering has numerous applications across industries, allowing AI to automate tasks, enhance productivity, and generate high-quality insights. These applications span across domains such as:

1. Business and Marketing – Customer service automation, content creation, and market analysis.
2. Education and Learning – AI tutoring, knowledge retrieval, and personalized study plans.

3. Healthcare and Medicine – Medical research summarization, diagnostic assistance, and patient education.
4. Programming and Software Development – Code generation, debugging, and software documentation.
5. Creative Writing and Content Creation – Generating stories, scripts, and articles.
6. Legal and Compliance – Contract analysis, policy generation, and legal research.
7. Scientific Research – Data summarization, literature reviews, and hypothesis generation.
8. Finance and Investment – Market predictions, risk analysis, and automated reporting.
9. AI Automation and Workflow Optimization – AI agents for repetitive tasks and business process automation.

Now, let us explore these applications in detail.

1. Business and Marketing Applications

Businesses leverage AI models through prompt engineering to automate tasks, analyze trends, and generate marketing content. Properly crafted prompts enable AI to provide:

Customer Support Automation

- AI chatbots and virtual assistants handle customer queries, complaints, and FAQs.
- Example Prompt: *"You are a customer support assistant for an e-commerce website. Provide a response to a customer asking for a refund due to a defective product."*

Marketing Content Creation

- AI generates compelling advertisements, blog posts, and email campaigns.
- Example Prompt: *"Write a persuasive Instagram ad for a new organic skincare product targeting young adults."*

Market Research and Competitive Analysis

- AI summarizes competitor strategies, industry trends, and customer sentiment.
- Example Prompt: *"Analyze recent consumer sentiment towards electric vehicles based on social media discussions and industry reports."*

2. Education and Learning

AI-powered tutoring and study aids help students and educators by providing personalized learning experiences.

AI Tutoring and Personalized Learning Plans

- AI acts as a tutor, explaining complex topics based on the student's learning level.
- Example Prompt: *"Explain the theory of relativity in simple terms for a high school student with examples."*

Automated Quiz and Exam Generation

- AI generates quizzes, tests, and study guides.
- Example Prompt: *"Create a 10-question multiple-choice quiz on World War II for middle school students."*

Summarizing Research Papers and Study Materials

- AI condenses long research papers and academic content into key takeaways.
- Example Prompt: *"Summarize the key findings of this research paper on artificial neural networks in 200 words."*

3. Healthcare and Medicine

AI supports medical professionals by analyzing research, assisting in diagnostics, and improving patient education.

Medical Research Summarization

- AI processes and summarizes complex medical studies.
- Example Prompt: *"Summarize the latest research on Alzheimer's disease treatment in non-technical language."*

Diagnostic Assistance and Symptom Analysis

AI helps doctors interpret symptoms and suggest possible conditions.

Example Prompt: *"Given the symptoms of persistent cough, fever, and difficulty breathing, suggest possible diagnoses and follow-up questions for a doctor."*

Patient Education and Health Advice

- AI explains medical conditions in simple terms.
- Example Prompt: *"Explain the causes and prevention methods of diabetes to a non-medical audience."*

4. Programming and Software Development

AI assists developers with coding, debugging, and software documentation.

Code Generation and Explanation

- AI writes code snippets based on user input.
- Example Prompt: *"Write a Python function that sorts a list of dictionaries by a specified key."*

Debugging and Error Resolution

- AI identifies and fixes coding errors.
- Example Prompt: *"Explain and correct the error in this Python function: [insert code]."*

Automated Software Documentation

- AI generates documentation for complex software systems.
- Example Prompt: *"Generate detailed API documentation for this Flask web application."*

5. Creative Writing and Content Creation

AI aids authors, screenwriters, and content creators in generating engaging content.

Story and Script Writing

- AI creates compelling narratives, plots, and character dialogues.
- Example Prompt: *"Write the opening scene of a sci-fi novel set in a dystopian future where AI controls the government."*

Blog and Article Writing

- AI produces informative and engaging articles.
- Example Prompt: *"Write a 1000-word article on the impact of artificial intelligence on modern journalism."*

Poetry and Lyrics Generation

- AI generates creative poetry and song lyrics.
- Example Prompt: *"Write a romantic poem in the style of Shakespeare about lost love."*

6. Legal and Compliance

Law firms and businesses use AI for contract analysis, compliance checks, and legal research.

Contract Review and Summarization

- AI identifies key clauses and obligations in legal agreements.
- Example Prompt: *"Summarize the main obligations and risks in this contract related to intellectual property rights."*

Legal Research and Case Law Analysis

- AI finds relevant legal precedents for court cases.
- Example Prompt: *"Provide a summary of U.S. Supreme Court rulings on digital privacy rights."*

7. Scientific Research

AI aids researchers by summarizing papers, analyzing data, and generating hypotheses.

Research Summarization

- AI condenses lengthy research studies into key points.
- Example Prompt: *"Summarize the latest advancements in quantum computing in layman's terms."*

Hypothesis Generation and Data Analysis

- AI proposes new research hypotheses based on existing data.
- Example Prompt: *"Based on recent trends in renewable energy research, suggest potential areas for further exploration."*

8. Finance and Investment

AI streamlines financial analysis, stock market predictions, and investment strategies.

Market Predictions and Trend Analysis

- AI analyzes stock market trends and economic indicators.
- Example Prompt: *"Summarize the key financial trends impacting the stock market in the last quarter."*

Portfolio Risk Assessment

- AI evaluates investment risks and opportunities.
- Example Prompt: *"Assess the risks and potential returns of investing in cryptocurrency versus traditional stocks."*

9. AI Automation and Workflow Optimization

Businesses integrate AI into workflows using prompt chaining and AI automation techniques.

Task Automation with AI Agents

- AI performs repetitive tasks such as email sorting, appointment scheduling, and data entry.
- Example Prompt: *"Automatically categorize and summarize incoming customer support emails."*

AI-Powered Business Intelligence

- AI generates reports and insights based on business data.
- Example Prompt: *"Analyze the sales data from the last three months and provide insights on seasonal trends."*

Conclusion

Prompt engineering significantly enhances AI capabilities, making it a powerful tool in various industries. By crafting precise prompts, users can:

- Improve efficiency and accuracy in AI-generated responses.
- Automate complex and repetitive tasks.
- Enhance decision-making processes across different fields.
- Drive innovation in business, healthcare, finance, and creative fields.

As AI continues to evolve, mastering prompt engineering will be essential for unlocking new possibilities in artificial intelligence.

CHAPTER 2

FUNDAMENTALS OF PROMPT DESIGN

Prompt design is the foundation of interacting effectively with artificial intelligence models. It determines how an AI system interprets input and generates responses. By understanding how AI processes prompts, different types of prompting techniques, and principles for effective prompting, users can guide AI models to produce more accurate, coherent, and useful outputs.

This section explores the fundamentals of prompt design in a highly detailed and structured manner.

Understanding How AI Models Process Prompts

Introduction

AI models, especially Large Language Models (LLMs) like GPT-4, Gemini, Claude, and LLaMA, are trained to understand and generate human-like text based on input prompts. However, they do not "think" or "understand" in the way humans do. Instead, they rely on statistical patterns learned from vast amounts of text data.

To effectively leverage AI for prompt engineering, it is crucial to understand how AI processes prompts, how it interprets context, and how it generates responses. This understanding enables users to craft precise and effective prompts to get the most accurate and relevant results.

1. The Basic Working Mechanism of AI Models

Step 1: Tokenization – Breaking Down the Input

When a prompt is entered into an AI model, the first thing it does is convert the text into tokens.

- Tokens are the fundamental units of text that the model processes.
- A token can be a word, part of a word, or even a character depending on the language model.
- Example:

  - The sentence *"AI is transforming the world."* might be tokenized as: ["AI", "is", "transform", "ing", "the", "world", "."]
- In some models, a whole word may be a single token, while in others, it may be split into smaller components.

Tokenization allows the AI to process each piece of text separately and understand the structure of a sentence in numerical form.

Step 2: Encoding – Converting Text into Vectors

After tokenization, the AI model converts the tokens into numerical representations known as embeddings.

- Embeddings are high-dimensional numerical vectors that represent words in a mathematical space.
- The model does not understand words as humans do but interprets relationships between words based on their numerical representations.
- Words with similar meanings or contexts are closer in the vector space, which helps the AI understand semantic relationships.

For example:

- "King" and "Queen" will have similar vector representations.
- "Paris" and "France" will be mathematically related in the AI's representation.

This process enables AI to capture the meaning of words in context rather than just recognizing individual words.

Step 3: Context Processing – Understanding the Prompt

Once the model has converted text into numerical data, it uses a Transformer architecture (like in GPT models) to process the input. The AI examines:

1. The order of words – Sentence structure affects meaning.
2. Contextual relationships – Words change meaning based on surrounding words.
3. Intent of the prompt – Whether the user is asking a question, requesting a summary, or seeking a creative response.

AI models process text using an attention mechanism called Self-Attention (from the Transformer model).

- The Self-Attention mechanism helps the AI focus on the most important words in the input.
- Example: In the sentence *"The cat sat on the mat because it was tired."*
  - The AI needs to determine what "it" refers to, which depends on the context of "cat" being the subject.

Step 4: Response Generation – Predicting the Next Word

Once the model understands the context, it predicts the next words based on probabilities.

- It does not "think" or "decide" like a human but selects words based on statistical likelihood.
- The AI generates responses one token at a time by choosing the most probable next token based on its training data.
- For instance, given the input "Once upon a time", the model might predict the next word as "there" because it frequently follows in fairy tales.

The quality of the response depends on:

- Prompt clarity – Vague prompts lead to vague responses.
- Length of the context – Shorter prompts provide less information for AI to generate precise responses.
- Temperature setting – A lower value (e.g., 0.2) makes responses more predictable, while a higher value (e.g., 0.8) makes them more creative and varied.

2. Factors Influencing AI's Response to Prompts

1. Context Window and Memory Limitations

AI models have a limited context window, meaning they can only process a fixed number of tokens at a time.

- If the prompt exceeds this limit, the model forgets earlier parts of the text.
- Example: If the context window is 4096 tokens, anything beyond that is discarded, affecting coherence in long-form content.

2. Temperature and Creativity

The temperature parameter controls the randomness of the model's output.

- Lower temperature (e.g., 0.2) → More predictable, factual responses.
- Higher temperature (e.g., 0.8) → More creative, varied responses.
- Example:
  - » Prompt: *"Suggest a futuristic city concept."*
  - » Temperature 0.2 → "A high-tech city with automated transportation."
  - » Temperature 0.8 → "A floating city above the ocean with AI-controlled weather systems and self-sustaining energy grids."

3. Prompt Length and Clarity

- Short and vague prompts often lead to generic responses.
- Longer and well-structured prompts help the AI understand intent better.

- Example:
  - » Poor Prompt: *"Tell me about space."*
  - » Better Prompt: *"Explain the process of star formation, including the role of gravity and nuclear fusion."*

4. Stop Sequences and Output Length

- AI generates responses until it reaches a predefined stop sequence (e.g., punctuation or special tokens).
- Users can set a max token limit to prevent excessively long outputs.

3. Best Practices for Effective Prompt Engineering

Understanding how AI models process prompts allows users to design better prompts that yield high-quality results.

1. Be Specific and Clear

- AI performs best when the prompt is detailed and well-defined.
- Example:
  - » Poor Prompt: *"Write about AI."*
  - » Better Prompt: *"Explain the impact of AI on healthcare, particularly in diagnostics and patient monitoring."*

2. Use Role-Based Prompting

- Assign a role to the AI to shape its response style.
- Example:
  - » *"You are an experienced data scientist. Explain machine learning algorithms to a beginner."*

3. Provide Examples for Context

- AI improves its responses when given examples.
- Example:
  - » *"Translate this sentence into Spanish: 'The weather is nice today.'"*

4. Use Step-by-Step Instructions

- AI follows multi-step prompts better.
- Example:
  - » *"Summarize this article in three bullet points, then provide a conclusion."*

5. Set the Format of the Response

- Specify the format, such as a list, table, paragraph, or structured output.
- Example:
  - » *"List five key benefits of renewable energy in bullet points."*

Conclusion

Understanding how AI processes prompts is essential for improving prompt engineering skills. AI models:

1. Tokenize and encode text into numerical representations.
2. Analyze context using transformer-based self-attention mechanisms.
3. Predict responses based on statistical probabilities.
4. Are influenced by prompt clarity, context length, temperature settings, and response formatting.

By applying this knowledge, users can optimize prompts to achieve more accurate, insightful, and relevant AI responses, ultimately improving efficiency and effectiveness in various AI-driven applications.

Types of Prompts in AI and Large Language Models

Introduction

Prompts serve as the primary way of interacting with AI models. They are the input instructions given to an AI system to generate a response. The effectiveness of an AI-generated output largely depends on the quality and structure of the prompt.

Different types of prompts are used based on the context, goal, and desired response. Understanding these types can help users design better prompts, improving accuracy and relevance in AI responses.

1. Instruction-Based Prompts

Definition

Instruction-based prompts give direct commands to the AI, specifying exactly what the user wants. These prompts are designed to provide clear guidance to the model about what to do.

Characteristics

- The AI is given an explicit instruction to follow.
- The response is often structured and predictable.
- Works well for task-oriented applications like summarization, translation, and answering specific questions.

Examples

1. **Summarization Prompt**

- *"Summarize the following article in three sentences."*

2. **Definition Prompt**

- *"Define the term 'blockchain' in simple words."*

3. **Translation Prompt**
- *"Translate the sentence 'Hello, how are you?' into French."*
4. **Comparison Prompt**
- *"Compare the differences between machine learning and deep learning."*

Use Cases
- Educational explanations
- Step-by-step instructions
- Factual responses
- Simple data retrieval tasks

2. Conversational Prompts

Definition

Conversational prompts are designed to make AI engage in human-like dialogue. Instead of a simple command, these prompts often simulate back-and-forth conversations, making AI responses more dynamic and interactive.

Characteristics
- More open-ended than instruction-based prompts.
- The AI responds in a natural, human-like manner.
- Useful for chatbots, virtual assistants, and customer service applications.

Examples
1. **Casual Conversation**
- *"Hi, how's your day going?"*
2. **Opinion-Based Question**
- *"What do you think about the future of artificial intelligence?"*
3. **Contextual Continuation**
- *User: "Tell me about space travel."*
- *AI: "Space travel involves using spacecraft to explore space. Would you like to know about past missions or future plans?"*

Use Cases
- AI chatbots and virtual assistants
- Customer support interactions
- Simulating human-like discussions
- Engaging storytelling and entertainment

3. Few-Shot Prompts

Definition

Few-shot prompting provides the AI with a few examples before asking it to generate a response. This technique helps the AI learn the desired response pattern from the examples given.

Characteristics

- Improves accuracy by showing AI the expected response style.
- Helps AI perform complex or nuanced tasks more effectively.
- More effective in models with limited prior knowledge on a subject.

Examples

1. **Text Classification**

- *"Classify the following movie reviews as positive or negative.*
    - » *Example 1: 'This movie was fantastic! I loved it.' – Positive*
    - » *Example 2: 'The film was boring and too long.' – Negative*
      Review: 'The plot was engaging, but the acting was weak.' – [Your Turn]"

2. **Grammar Correction**

- *"Fix grammatical errors in these sentences.*
    - » *Example: 'She go to the market.' → 'She goes to the market.'*
    - » *Sentence: 'He like football.' → [Your Turn]"*

Use Cases

- Training AI to follow a specific format.
- Improving classification tasks.
- Enhancing complex problem-solving.

4. Zero-Shot Prompts

Definition

Zero-shot prompting asks the AI to perform a task without providing any prior examples. It relies on the AI's pre-existing knowledge to generate responses.

Characteristics

- Works well when the AI has strong training data on the topic.
- Can lead to less accurate results if the task requires nuanced understanding.
- Useful when examples are not available or necessary.

Examples

1. **General Question**
- *"What is quantum computing?"*
2. **Summarization Task**
- *"Summarize this article in one paragraph."*
3. **Sentence Completion**
- *"Complete the sentence: 'Artificial intelligence is changing the world by...'"*

Use Cases

- When AI has good prior knowledge of the topic.
- General queries and factual explanations.
- Rapid brainstorming or idea generation.

5. Role-Based Prompts

Definition

Role-based prompts instruct the AI to take on a specific persona or expertise level before generating a response. This helps in obtaining responses tailored to a particular style, tone, or domain expertise.

Characteristics

- AI adapts its language and tone based on the assigned role.
- More specialized and industry-specific responses.
- Helps create engaging, context-aware conversations.

Examples

1. **Teacher Role**
- *"You are a history professor. Explain the causes of World War II in simple terms."*
2. **Software Engineer Role**
- *"You are a senior software developer. Explain how recursion works in Python."*
3. **Doctor Role**
- *"You are a medical expert. Explain the symptoms of diabetes and possible treatments."*

Use Cases

- Simulating expert opinions.
- Enhancing realism in AI-generated characters.
- Providing customized learning experiences.

6. Structured Formatting Prompts

Definition

These prompts instruct the AI to return responses in a specific format, such as tables, bullet points, JSON, or step-by-step instructions.

Characteristics

- Makes responses easier to read and interpret.
- Ensures AI-generated content is structured and organized.
- Useful for data extraction, automation, and programming tasks.

Examples

1. **Bullet Point List**
- *"List the key benefits of renewable energy in bullet points."*
2. **Table Format**
- *"Provide a table comparing electric cars and gasoline cars based on cost, efficiency, and environmental impact."*
3. **JSON Output**
- *"Return the following data in JSON format: Name, Age, Country."*

Use Cases

- Automating content generation.
- Structuring AI responses for databases or APIs.
- Ensuring AI-generated data is machine-readable.

Conclusion

Understanding the different types of prompts is essential for optimizing AI interactions. Each type serves a specific purpose, from giving direct instructions to engaging in natural conversations, providing examples, or structuring responses in a readable format.

By selecting the appropriate prompt type, users can achieve more accurate, meaningful, and effective AI outputs, improving the overall efficiency of AI-assisted tasks in various domains.

Principles of Effective Prompting

Introduction

Prompting is the technique of designing input instructions for an artificial intelligence model to generate the most accurate, relevant, and useful response. The effectiveness of artificial intelligence generated content depends heavily on how well the prompt is structured. Poorly designed prompts can lead to vague, irrelevant, or incorrect responses.

To ensure optimal results, it is essential to follow specific principles of effective prompting. These principles help improve clarity, specificity, consistency, and contextual understanding, enabling artificial intelligence to generate high quality responses.

Clarity and Precision

A well designed prompt should be clear, precise, and unambiguous. The artificial intelligence model needs to understand exactly what is being asked to provide an accurate response.

A clear and precise prompt reduces the chances of artificial intelligence misinterpreting the request, leading to more accurate and relevant answers. It avoids unnecessary assumptions by the artificial intelligence model.

To achieve clarity and precision, use simple and direct language, avoid vague or ambiguous terms, and be explicit about what is expected from the artificial intelligence.

For example, a prompt asking to tell about artificial intelligence is too vague. Instead, a prompt asking for a detailed explanation of artificial intelligence, including its definition, history, and key applications, is more precise and effective.

Specificity and Context

Providing context and specific details helps artificial intelligence generate more focused and relevant responses. Without proper context, artificial intelligence may produce generic or inaccurate answers.

A specific and contextualized prompt helps artificial intelligence better understand the intent behind the request, reducing unnecessary guesswork and ensuring the response matches the user's expectations.

To ensure specificity and context, include necessary background information, specify the desired format such as a summary, explanation, or step by step guide, and use role based prompting if needed.

For example, instead of simply asking for an explanation of machine learning, a better prompt would specify the context, such as explaining machine learning in healthcare and focusing on how it helps in disease prediction and diagnosis.

Providing Examples for Guidance

Including examples within the prompt helps artificial intelligence understand the desired response format, tone, and style. This technique is especially useful in few shot learning, where artificial intelligence is shown examples before generating a response.

Providing examples helps artificial intelligence align its responses with user expectations, improves accuracy, and ensures consistency in responses.

To use this principle effectively, provide one or more examples to guide artificial intelligence, show the desired output style explicitly, and ensure examples are clear and relevant to the prompt.

For instance, instead of just asking for a summary of an article, a prompt should specify that the summary should be in three bullet points, each containing a key point. Similarly, when classifying reviews, an example of a positive and a negative review should be included to guide artificial intelligence.

Structuring Prompts for Better Readability

Breaking down complex instructions into well structured prompts makes it easier for artificial intelligence to understand and respond effectively.

A structured prompt improves artificial intelligence's ability to process multi step requests, reduces the risk of misinterpretation, and ensures the output follows a logical order.

To enhance readability, use bullet points or numbered lists for clarity, separate multiple instructions clearly, and indicate preferred formatting such as paragraphs, tables, or lists.

For example, instead of asking for an explanation of Newton's laws of motion in an unstructured manner, a better prompt would request an explanation in a structured format, asking artificial intelligence to state the law, provide a simple explanation, and give a real world example.

Using Constraints and Output Formatting

Constraints help control artificial intelligence's response in terms of length, style, and content structure.

Applying constraints prevents overly long or too short responses, helps maintain consistent style and structure, and ensures artificial intelligence generated content is readable and user friendly.

To use this principle effectively, set word limits if necessary, specify tone and style such as formal, casual, or scientific, and use output formatting instructions.

For instance, instead of a general request for summarizing an article, specifying that the summary should be exactly one hundred words and use concise language makes the output more structured and useful. Similarly, when describing the benefits of meditation, requesting a list of five key benefits in a formal, research based tone ensures clarity and precision.

Iterative Prompt Refinement

Refining prompts based on artificial intelligence generated responses helps improve their effectiveness. If the response is unclear, vague, or off topic, modifying the prompt can optimize results.

Iterative refinement helps improve accuracy and relevance, reduces misinterpretations and inconsistencies, and ensures artificial intelligence learns the desired response pattern over time.

To refine prompts effectively, test different variations of a prompt, adjust wording, add context, or provide examples, and use feedback loops to refine and improve the prompt.

For example, if a prompt asking about the effects of caffeine results in a generic response, refining the prompt to specify both positive and negative effects, focusing on cognitive function, energy levels, and long term consumption risks, leads to a more precise and comprehensive response.

Conclusion

Effective prompting is essential for obtaining accurate, relevant, and well structured responses from artificial intelligence. By applying principles such as clarity, specificity, structured formatting, examples, constraints, and iterative refinement, users can significantly improve artificial intelligence generated results.

Mastering these principles allows users to harness the full potential of artificial intelligence models, ensuring that interactions are efficient, productive, and aligned with user expectations.

CHAPTER 3

BASIC PROMPTING TECHNIQUES

Prompting techniques are essential for effectively interacting with large language models (LLMs) such as GPT-4. Understanding how to structure prompts correctly can significantly impact the quality and relevance of the responses.

This section covers the fundamental prompting techniques, including:

1. Direct Instructions
2. Role-Based Prompting
3. Formatting and Structuring Prompts
4. Adjusting Prompt Specificity and Clarity

Each technique plays a crucial role in guiding AI to produce more accurate, relevant, and context-aware responses.

Direct Instructions

Direct instruction prompting is the most straightforward way to interact with an AI model. It involves clearly stating what the user wants the AI to do.

How Direct Instructions Work

- AI models generate responses based on patterns in their training data.
- The more precise and direct an instruction is, the better the AI can understand and execute the task.
- Ambiguous or vague instructions can lead to incorrect or incomplete responses.

Examples of Direct Instructions

Basic Direct Instructions

These prompts contain clear commands, ensuring that AI understands the user's intent.

- Prompt: "List five benefits of regular exercise."

- Response:

1. Improves cardiovascular health
2. Enhances mental well-being
3. Helps in weight management
4. Strengthens muscles and bones
5. Boosts energy levels

Step-by-Step Instructions

When a task requires multiple steps, structuring the prompt as a step-by-step guide ensures clarity.

- Prompt: "Explain how to change a car tire in step-by-step instructions."
- Response:

1. Park the car on a flat surface and engage the parking brake.
2. Use a wrench to loosen the lug nuts slightly.
3. Place a jack under the vehicle and lift it until the tire is off the ground.
4. Remove the lug nuts completely and take off the flat tire.
5. Place the spare tire on the wheel hub and hand-tighten the lug nuts.
6. Lower the car and tighten the lug nuts securely with a wrench.

Direct Instructions with Formatting Requests

You can instruct AI to format responses in a specific way.

- Prompt: "Summarize the causes of World War One in a table format with two columns: Cause and Explanation."
- Response:

Cause	Explanation
Militarism	Nations built large armies, increasing tensions.
Alliances	Secret alliances led to widespread involvement in the war.
Imperialism	Competition for colonies created conflicts between nations.
Nationalism	Intense national pride fueled aggressive policies.

Best Practices for Direct Instructions

1. Use Command Words: Start prompts with words like "List," "Explain," "Describe," "Summarize," or "Compare" to make the instruction clear.
2. Be Specific: Instead of "Tell me about space," say, "Explain three key discoveries made by the Hubble Space Telescope."
3. Provide Constraints: Specify word limits, formatting preferences, or output style if necessary.

Role-Based Prompting

Role-based prompting involves instructing AI to assume a specific role or perspective when generating a response. This technique enhances contextual relevance and improves the quality of responses for professional, technical, or creative tasks.

How Role-Based Prompting Works

- AI models adjust their language, tone, and style based on the assigned role.
- This is useful in industries such as education, customer service, healthcare, law, and creative writing.

Examples of Role-Based Prompting

Professional Expertise Simulation

- Prompt: "You are a medical doctor. Explain the symptoms and treatment options for migraine headaches."
- Response: "As a doctor, I can explain that migraines are neurological conditions characterized by severe headaches, nausea, and sensitivity to light and sound. Treatment includes pain relievers, preventive medications, and lifestyle changes such as avoiding triggers and managing stress."

Customer Support Role

- Prompt: "You are a customer service representative for a software company. A user is frustrated about a billing issue. How would you respond?"
- Response: "I understand how frustrating billing issues can be. Let me assist you in resolving this matter. Could you provide your account details so I can review the charges? If an error is found, we will correct it immediately. I appreciate your patience."

Creative Role-Based Prompting

- Prompt: "You are William Shakespeare. Write a short poetic description of the moon."
- Response: "Lo! The silver sphere dost wax and wane, A lantern hung in heaven's domain, Softly kissing the midnight sky, A lover's muse and dreamer's eye."

Best Practices for Role-Based Prompting

1. Clearly Define the Role: Start the prompt with "You are a…" to establish the perspective.
2. Provide Context: If the role requires specialized knowledge, specify the topic or scenario.
3. Use Role-Based Tone: Encourage AI to maintain the tone and expertise level of the assigned role.

Formatting and Structuring Prompts

The way a prompt is formatted influences the AI's response. Well-structured prompts lead to clear, concise, and organized answers.

Importance of Formatting in Prompting

- Helps AI understand the structure of the response.
- Makes the output more readable and user-friendly.
- Prevents misunderstandings or incomplete answers.

Examples of Formatting in Prompts

List Format

- Prompt: "List three major causes of deforestation."
- Response:

1. Agricultural expansion
2. Logging for timber
3. Urban development

Table Format

- Prompt: "Compare renewable and non-renewable energy sources in a table format."
- Response:

Type	Examples	Pros	Cons
Renewable	Solar, Wind, Hydro	Sustainable, Low Emissions	High Initial Cost
Non-renewable	Coal, Oil, Gas	High Energy Output	Pollution, Limited Supply

Bullet Points

- Prompt: "Describe the effects of climate change in bullet points."
- Response:
  - » Rising global temperatures
  - » Increased frequency of natural disasters
  - » Rising sea levels due to polar ice melting

Best Practices for Formatting Prompts

1. Use Keywords like 'List,' 'Table,' or 'Bullet Points' to guide AI.
2. Provide Clear Examples if necessary.
3. Specify the Number of Items to control response length.

Adjusting Prompt Specificity and Clarity

Why Prompt Specificity Matters

- Broad prompts lead to vague responses.
- Highly specific prompts guide AI to generate more useful outputs.

Examples of Adjusting Prompt Specificity

Vague vs. Specific Prompts

- Vague: "Tell me about space."
- Specific: "Explain how black holes form and describe their gravitational effects on nearby objects."
- Vague: "Summarize the book *1984*."
- Specific: "Summarize the main themes of *1984* with examples from the text."

Techniques to Improve Clarity

1. Avoid Ambiguity: Instead of "Tell me about Paris," specify "Describe the history, culture, and famous landmarks of Paris."
2. Set Constraints: Define word limits or output formats.
3. Use Follow-Up Questions: If needed, refine prompts based on the initial AI response.

Conclusion

Mastering basic prompting techniques improves AI interactions. By using direct instructions, role-based prompting, formatting structures, and adjusting specificity, users can enhance the clarity, accuracy, and usefulness of AI-generated responses.

CHAPTER 4
INTERMEDIATE PROMPTING STRATEGIES

As artificial intelligence models become more advanced, effective prompting strategies help improve the accuracy, relevance, and depth of responses. Intermediate prompting techniques enhance AI's reasoning, adaptability, and ability to handle complex tasks. This section explores chain-of-thought prompting, few-shot learning, zero-shot versus few-shot prompting, and prompt chaining with context retention.

What is Chain-of-Thought Prompting?

Chain-of-thought prompting is a prompting technique that guides an artificial intelligence model to break down its reasoning process step by step, rather than providing a direct answer. This method helps improve the AI's ability to handle complex reasoning tasks, multi-step problems, and logical decision-making processes.

This technique is especially useful for large language models because, unlike traditional programming-based AI systems, they do not inherently have structured logic or reasoning built-in. Instead, they predict text based on patterns in data. By explicitly instructing the model to outline its reasoning, we can enhance its ability to make correct and interpretable decisions.

How Chain-of-Thought Prompting Works

AI models, including large language models like ChatGPT, generate responses based on probability distributions learned from training data. When given a question, they predict the most likely response based on prior examples. However, for complex questions that require reasoning or calculations, the model may not always arrive at the best answer.

By structuring prompts using chain-of-thought methodology, we enable the AI to break down the problem, analyze different aspects step by step, and reach a well-thought-out conclusion.

For example, instead of simply asking: "What is 24 multiplied by 17?"

A chain-of-thought prompt would be: "Solve 24 multiplied by 17 step by step, explaining each step of the calculation."

This forces the AI to break the problem down into intermediate steps, improving accuracy.

Benefits of Chain-of-Thought Prompting

1. Improves Accuracy in Complex Problems

- AI models often struggle with mathematical or logical problems when asked to provide direct answers. CoT prompting helps improve accuracy by making the model think through its response logically.

2. Enhances Interpretability

- When AI explains its reasoning, users can understand how it arrived at the answer. This makes it easier to verify and trust the response.

3. Reduces Errors in Multi-step Reasoning Tasks

- Many errors in AI-generated responses stem from skipping intermediate steps. CoT prompting ensures that each step is considered, reducing the likelihood of mistakes.

4. Encourages Logical and Critical Thinking

- Instead of responding based purely on probability, the model is encouraged to apply structured logic, which is beneficial for subjects like mathematics, coding, and scientific reasoning.

5. Helps in Educational and Training Scenarios

- In tutoring or e-learning applications, chain-of-thought prompting allows AI to act as a tutor, explaining concepts rather than just providing answers.

Examples of Chain-of-Thought Prompting

Example 1: Basic Arithmetic Without CoT Prompting

Prompt: "What is 37 plus 48?" AI Response: "The answer is 85."

While the answer is correct, there is no explanation. If the AI had made a mistake, the user would not know where the error occurred.

Example 2: Basic Arithmetic With CoT Prompting

Prompt: "Solve 37 plus 48 step by step, explaining each step clearly." AI Response:

1. Break the numbers into place values:

- 37 consists of 30 and 7.
- 48 consists of 40 and 8.

2. Add the tens first:

- 30 plus 40 equals 70.

3. Add the ones next:
- 7 plus 8 equals 15.
4. Combine both results:
- 70 plus 15 equals 85.
5. The final answer is 85.

By explicitly detailing each step, the AI ensures a more structured and understandable response.

Example 3: Logical Reasoning Without CoT Prompting

Prompt: "John is taller than Mark. Mark is taller than Alice. Who is the shortest?" AI Response: "Alice is the shortest."

While this is correct, it does not show how the AI reached this conclusion.

Example 4: Logical Reasoning With CoT Prompting

Prompt: "John is taller than Mark. Mark is taller than Alice. Explain step by step who is the shortest." AI Response:

1. We know that John is taller than Mark.
2. We also know that Mark is taller than Alice.
3. Since Alice is not taller than either John or Mark, Alice must be the shortest.

By breaking the reasoning into steps, the AI makes the logic easier to follow and verify.

Advanced Applications of Chain-of-Thought Prompting

1. Solving Math and Science Problems

CoT prompting is particularly useful in STEM fields where complex multi-step calculations are involved.

Example: Prompt: "A car travels 60 kilometers per hour. How far does it travel in three and a half hours? Solve step by step." AI Response:

1. Identify the given values:
- Speed of the car = 60 km/h
- Time traveled = 3.5 hours
2. Use the formula:
- Distance = Speed Time
3. Multiply the values:
- Distance = 60 3.5
4. Break down the multiplication:
- 60 3 = 180

- 60 0.5 = 30
- Total distance = 180 + 30 = 210 kilometers

5. Final answer: 210 kilometers.

2. Improving AI Coding Assistance

CoT prompting is effective in coding-related tasks where step-by-step breakdowns help prevent errors.

Example: Prompt: "Write a Python function to check if a number is prime. Explain each step of your code." AI Response:

1. Define a function is_prime(n).
2. If n is less than 2, return False because prime numbers are greater than 1.
3. Check divisibility from 2 to sqrt(n).
4. If any number divides n, return False.
5. If no divisors are found, return True.

python

CopyEdit

```
import math
def is_prime(n):
    if n < 2:
        return False
    for i in range(2, int(math.sqrt(n)) + 1):
        if n % i == 0:
            return False
    return True
```

By explaining each step, the AI provides a clearer, more educational response.

3. Decision-Making and Analysis

CoT prompting is valuable in business and finance, where structured reasoning helps evaluate different choices.

Example: Prompt: "A company is considering expanding into a new market. What factors should they evaluate? Explain step by step." AI Response:

1. Market Demand – Analyze if there is enough demand for the company's products in the new market.
2. Competition – Identify existing competitors and their strengths and weaknesses.
3. Regulations – Understand local laws and industry regulations.

4. Costs and Profitability – Estimate costs and expected revenue to ensure profitability.
5. Logistics and Supply Chain – Assess the feasibility of supplying products to the new market.
6. Brand Fit – Determine if the company's brand aligns with local consumer preferences.

Conclusion

Chain-of-thought prompting is a powerful technique that enhances AI responses by breaking down complex reasoning into logical steps. It improves accuracy, interpretability, and error reduction, making it particularly useful for mathematics, science, coding, and decision-making tasks.

Key Takeaways:

CoT prompting helps AI think step by step, improving response accuracy.

It is especially useful for logical reasoning, mathematics, and multi-step tasks.

Breaking down complex tasks into intermediate steps enhances clarity and reliability.

Applications include education, coding, business analysis, and technical explanations.

Few-shot Learning

What is Few-shot Learning?

Few-shot learning is a technique where AI is given multiple examples before being asked to generate a response. This helps the model learn patterns and produce more accurate, contextual outputs.

Why is Few-shot Learning Effective?

- Provides AI with examples that shape its response style.
- Helps improve accuracy in tasks requiring specialized formatting or structured answers.
- Makes AI adapt to specific writing styles, tones, and problem-solving methods.

Example of Few-shot Learning

Without Few-shot Learning

Prompt: "Translate the phrase 'How are you?' into French." **AI Response:** "Comment ça va?"

This works, but for more complex tasks, the AI may not provide responses in the correct format.

With Few-shot Learning

Prompt: "Translate the following English sentences into French:

1. Hello, my name is John. → Bonjour, je m'appelle John.
2. Where is the nearest train station? → Où est la gare la plus proche?
3. How are you? →"

AI Response: "Comment ça va?"

By providing examples first, the AI follows the given structure, ensuring accuracy and consistency.

Applications of Few-shot Learning

- **Language Translation** – Improves translation quality by following provided examples.
- **Programming Assistance** – Guides AI to produce code in a specific style or format.
- **Creative Writing** – Helps AI match the tone of a given narrative or poetic style.
- **Text Summarization** – Allows AI to match the summary style to given examples.

Zero-shot vs. Few-shot Prompting

What is Zero-shot Prompting?

Zero-shot prompting is when the AI is given a task without any examples or prior guidance. The AI relies solely on its pre-trained knowledge to generate a response.

Example of Zero-shot Prompting

Prompt: "Write a short email declining a job offer politely." **AI Response:** "Dear Hiring Manager,

Thank you for offering me the [Job Title] position at [Company Name]. I appreciate the opportunity, but I have decided to pursue another opportunity. I am grateful for your time and consideration.

Best regards, [Your Name]"

Zero-shot prompting is useful when the AI has enough background knowledge on a topic. However, the response may lack personalization or structure if the task is complex.

What is Few-shot Prompting?

Few-shot prompting provides the AI with examples before asking it to generate a response. This is useful when responses require a specific format, style, or detailed knowledge.

Example of Few-shot Prompting

Prompt: "Here are two examples of polite job offer rejection emails:

1. Dear Hiring Manager, Thank you for considering me for the [Job Title] role. After careful thought, I have decided to decline the offer. I appreciate the opportunity and wish your team continued success.
2. Dear [Recruiter's Name], I am grateful for the opportunity to join [Company Name]. However, after careful consideration, I have decided to decline the offer. Thank you for your time and understanding.

Now, write a similar email politely declining a job offer."

The AI will generate a response that closely follows the structure and tone of the provided examples.

When to Use Zero-shot vs. Few-shot Prompting

- **Use Zero-shot Prompting When:**
  - » The task is simple, such as defining terms or answering factual questions.
  - » The AI has sufficient pre-trained knowledge on the topic.
  - » Quick and general responses are needed.
- **Use Few-shot Prompting When:**
  - » The response requires a specific format or style.
  - » The task involves complex reasoning or nuanced writing.
  - » Higher accuracy and structure are needed.

Prompt Chaining and Context Retention

What is Prompt Chaining?

Prompt chaining is the technique of breaking down complex tasks into multiple prompts, where each response builds upon the previous one. This helps AI maintain context across multiple turns and refine responses progressively.

Why is Prompt Chaining Useful?

- Helps maintain context over multiple interactions.
- Breaks down complex topics into step-by-step responses.
- Improves accuracy and relevance in longer conversations.

Example of Prompt Chaining

Without Prompt Chaining

Prompt: "Explain artificial intelligence and its applications in one response."
AI Response: "Artificial intelligence is the simulation of human intelligence in machines. It has applications in healthcare, finance, education, and more."

This response lacks depth and structure.

With Prompt Chaining

Prompt 1: "Explain artificial intelligence in simple terms." **AI Response:** "Artificial intelligence is a branch of computer science that enables machines to learn and perform tasks that typically require human intelligence, such as recognizing speech or making decisions."

Prompt 2: "What are the main applications of artificial intelligence?" **AI Response:** "Artificial intelligence is used in healthcare for diagnosing diseases, in finance for fraud detection, in education for personalized learning, and in transportation for self-driving cars."

By chaining prompts, AI provides a clearer and more detailed explanation.

Applications of Prompt Chaining

- **Storytelling and Content Generation** – Helps AI develop narratives progressively.
- **Research Assistance** – Allows AI to refine responses through multiple queries.
- **Technical Explanations** – Breaks down complex concepts into structured steps.
- **Customer Support** – Helps AI handle multi-step troubleshooting inquiries.

Conclusion

Intermediate prompting strategies significantly enhance AI interactions by improving reasoning, accuracy, and context retention. By using chain-of-thought prompting, few-shot learning, structured examples, and prompt chaining, users can refine AI-generated responses for better quality and effectiveness.

CHAPTER 5

ADVANCED PROMPT ENGINEERING

As artificial intelligence (AI) and large language models (LLMs) continue to evolve, more sophisticated prompting techniques are necessary to achieve high-quality, consistent, and structured outputs. Advanced prompt engineering strategies go beyond basic and intermediate techniques by introducing self-consistency prompting, tree-of-thought prompting, program-aided language models (PAL), and augmenting prompts with external tools. These techniques help improve accuracy, logical reasoning, and overall performance in complex tasks.

Self-Consistency Prompting

What is Self-Consistency Prompting?

Self-consistency prompting is an advanced technique in prompt engineering designed to enhance the reliability, accuracy, and consistency of AI-generated responses. Instead of relying on a single response, this method prompts the AI to generate multiple responses to the same question and then selects the most common or logical answer.

This technique is particularly useful for reducing randomness, improving reasoning, and ensuring more accurate outputs, especially in complex tasks such as mathematical problem-solving, logical reasoning, and decision-making.

Why is Self-Consistency Prompting Important?

Large language models (LLMs) generate responses probabilistically. This means that even with the same input, the AI may produce different answers each time due to variations in token probabilities. While this is useful for creativity, it can lead to inconsistent results in tasks that require logical precision.

Self-consistency prompting addresses this issue by:

1. Reducing Variability: Ensures more stable and repeatable responses.
2. Enhancing Logical Accuracy: Helps AI reach the most rational conclusion.
3. Filtering Out Errors: Eliminates incorrect or misleading responses by choosing the most common correct answer.

4. Mimicking Human Decision-Making: Similar to human decision-making, where multiple perspectives or trials help reinforce correct answers.

How Does Self-Consistency Prompting Work?

Self-consistency prompting involves three main steps:

1. Generating Multiple Responses

- Instead of taking a single output, the AI is prompted multiple times for the same question.
- Each response might differ slightly in reasoning, structure, or conclusion.

2. Analyzing and Comparing Responses

- The different outputs are examined to identify common patterns or the most frequently occurring answer.
- If an answer appears consistently across different responses, it is likely the most reliable.

3. Selecting the Most Consistent Answer

- The AI selects the response that appears the most or follows the most logical pathway.
- This ensures that the final output is more accurate and reliable than any single response.

Example of Self-Consistency Prompting

Scenario: Solving a Logical Puzzle

Prompt: *"A farmer is taking a wolf, a goat, and a cabbage across a river. He can only take one at a time in the boat. If left alone together, the wolf will eat the goat, or the goat will eat the cabbage. How can he transport all safely?"*

Without Self-Consistency Prompting (One Response)

- Response 1: "Take the wolf first, then the goat, then the cabbage." *(Incorrect; the goat will eat the cabbage.)*

With Self-Consistency Prompting (Multiple Responses)

- Response 1: "Take the goat first, then the wolf, then return with the goat, take the cabbage, and finally bring the goat again." *(Correct solution.)*
- Response 2: "Take the goat, then the cabbage, then the wolf." *(Incorrect; the wolf will eat the goat.)*
- Response 3: "First take the goat, then bring the wolf, but take the goat back. Then take the cabbage, and finally bring the goat again." *(Correct solution.)*
- Final Answer (Most Consistent Response Chosen): "Take the goat first, then the wolf, then return with the goat, take the cabbage, and finally bring the goat again."

By analyzing multiple responses, the AI determines the best approach and eliminates incorrect answers.

Use Cases of Self-Consistency Prompting

1. Mathematics and Logical Reasoning

- When solving equations or logical problems, AI may generate different steps or solutions. Self-consistency prompting ensures the AI selects the most frequent correct approach.
- Example:
  - » Prompt: "What is the square root of 144?"
  - » Responses: "12", "11.98", "12.0001"
  - » Final Answer: "12" (most frequent and accurate response).

2. Complex Decision-Making

- Used in business and strategic planning when multiple scenarios exist.
- Example: Choosing the best marketing strategy based on multiple AI-generated options.

3. Fact-Based Queries

- Ensures accurate responses by comparing multiple versions of AI-generated facts.
- Example: AI checks multiple responses to verify historical events or scientific facts.

4. Medical and Legal Advice

- Self-consistency prompting can help AI verify medical symptoms, diagnoses, or legal interpretations by filtering out incorrect suggestions.

Best Practices for Self-Consistency Prompting

1. Use in High-Stakes Tasks: Apply this technique for logical reasoning, decision-making, and factual accuracy.
2. Set a Defined Number of Responses: Ask the AI to generate multiple answers (e.g., five or ten responses) to improve consistency.
3. Compare Answer Patterns: Identify the most frequent or logically sound response.
4. Apply to Multi-Step Problems: Use for tasks requiring multiple steps where small errors can accumulate.

Limitations of Self-Consistency Prompting

1. Computational Cost: Generating multiple responses increases processing time and resource usage.

2. Not Always Needed: In simple tasks, self-consistency prompting may be unnecessary and redundant.
3. Requires Further Filtering: Sometimes, a consistent answer may still be incorrect if AI's logic is flawed.

Conclusion

Self-consistency prompting is a powerful technique for improving reliability, logical reasoning, and accuracy in AI-generated responses. By generating multiple answers, identifying patterns, and selecting the most consistent response, this approach reduces variability and enhances the overall performance of large language models.

Tree-of-Thought Prompting

What is Tree-of-Thought Prompting?

Tree-of-Thought (ToT) prompting is an advanced AI prompting technique that enhances problem-solving by breaking down complex tasks into multiple branches of reasoning, similar to how a decision tree works. Instead of generating a single, linear response, the AI explores multiple possible solutions, evaluates them step by step, and selects the most logical path.

This technique mimics human-like reasoning, where individuals break down problems into smaller components, explore different solutions, and choose the best approach.

Why is Tree-of-Thought Prompting Important?

Most traditional AI models generate answers in a single-pass format, meaning they respond immediately based on pattern recognition rather than structured reasoning. While effective for simple queries, this can lead to flawed logic, incomplete answers, or a lack of deep reasoning when handling complex problems.

Key Benefits of Tree-of-Thought Prompting:

1. Enhances Logical Reasoning: AI explores different thought processes before settling on an answer.
2. Improves Multi-Step Problem Solving: Ideal for complex reasoning tasks, mathematical problems, and strategic planning.
3. Reduces Errors: By analyzing multiple pathways, AI can eliminate incorrect or weak solutions.
4. Better Decision-Making: Helps AI evaluate multiple perspectives before finalizing a response.

How Does Tree-of-Thought Prompting Work?

Tree-of-Thought prompting follows a structured, multi-step approach to reach an optimized solution:

1. Breaking Down the Problem into Subproblems

- The AI does not immediately generate a final answer.
- Instead, it identifies key components of the problem and explores different pathways.

2. Exploring Multiple Thought Paths (Branching)

- Each subproblem has different possible solutions.
- The AI generates multiple possible approaches instead of committing to one immediately.

3. Evaluating and Selecting the Best Solution

- AI compares all the solutions and selects the most logical, accurate, or optimal answer.
- This ensures the response is well-reasoned rather than being a quick guess.

Example of Tree-of-Thought Prompting

Scenario 1: Solving a Logic Puzzle

Problem Statement: *"A man is looking at a picture of someone. His friend asks, 'Who is it?' The man replies, 'Brothers and sisters, I have none. But that man's father is my father's son.' Who is in the picture?"*

Without Tree-of-Thought Prompting (One-Pass Response)

- AI might quickly respond: "It's the man's father." *(Incorrect answer due to misinterpretation.)*

With Tree-of-Thought Prompting (Multi-Step Reasoning)

1. **Step 1: Understanding the Statement**

- "Brothers and sisters, I have none." → The speaker is an only child.
- "That man's father is my father's son."

2. **Step 2: Breaking Down the Sentence**

- "My father's son" → Since the speaker has no siblings, "my father's son" refers to himself.
- "That man's father is me." → The speaker is the father of the person in the picture.

3. **Step 3: Final Answer**

- The picture is of his son.

By structuring the reasoning process in a step-by-step, tree-like approach, the AI ensures that it interprets the logic correctly and eliminates incorrect conclusions.

Scenario 2: Solving a Chess Move Using Tree-of-Thought Prompting

Problem Statement: *"You are playing as White in a chess game. Your opponent's king is exposed. What is the best move to checkmate in two moves?"*

Tree-of-Thought Approach

1. Step 1: Identify Possible Checkmate Paths
- Examine different piece movements (e.g., queen, rook, bishop, knight).
- Consider defensive moves by the opponent.
2. Step 2: Branching Out Possible Moves
- Move 1A: Queen to h6 (puts king in check).
- Move 1B: Rook to g8 (forces king to move).
- Move 1C: Bishop to f5 (adds additional threat).
3. Step 3: Evaluating Opponent's Responses
- If Queen h6 → Opponent's king has only one escape square.
- If Rook g8 → Opponent has a forced move that leads to checkmate.
4. Step 4: Selecting the Best Move
- AI selects the best move based on likelihood of success → Rook to g8 leads to forced checkmate in two moves.

By considering multiple move options and evaluating opponent responses, AI makes a strategic decision rather than guessing the first available move.

Applications of Tree-of-Thought Prompting

1. Mathematical Problem Solving
- Used for multi-step algebra, calculus, and logical proofs.
- Example:
  - » Problem: "Find x if 2x + 3 = 15."
  - » Tree-of-Thought Breakdown:
    - ◆ Step 1: Subtract 3 from both sides → 2x = 12.
    - ◆ Step 2: Divide by 2 → x = 6.
2. **Scientific Analysis**
- Useful for AI in research fields like physics, chemistry, and biology.
- Example:
  - » Problem: "Describe the steps in DNA replication."
  - » Tree-of-Thought Approach:
    - ◆ Step 1: Initiation → Helicase unwinds the DNA strand.
    - ◆ Step 2: Elongation → DNA polymerase adds nucleotides.
    - ◆ Step 3: Termination → Ligase seals fragments.

3. **Business and Strategic Decision-Making**

- Helps AI analyze multiple business strategies before making recommendations.
- Example:
  - » Problem: "Which marketing strategy is best for increasing online sales?"
  - » Tree-of-Thought Breakdown:
    - ◆ Option 1: Increase social media ads.
    - ◆ Option 2: Improve website SEO.
    - ◆ Option 3: Offer discounts and loyalty programs.
    - ◆ AI evaluates customer behavior and selects the best approach.

4. **AI in Gaming and Robotics**

- Used in AI-powered chess engines, self-driving cars, and robotic decision-making.
- Example:
  - » Self-driving car deciding a route:
    - ◆ Option 1: Take Route A (faster but high traffic).
    - ◆ Option 2: Take Route B (slower but safer).
    - ◆ AI calculates and picks the best option dynamically.

Best Practices for Tree-of-Thought Prompting

1. Encourage Multi-Step Thinking:

- Instead of asking for a single answer, structure prompts to encourage step-by-step solutions.

2. Ask AI to Justify Answers:

- Prompt: *"Explain your reasoning step by step before finalizing an answer."*

3. Use for Open-Ended and Complex Problems:

- Works best for tasks that require deep analysis and structured reasoning.

4. Limit Overly Broad Trees:

- Too many branches can make AI overcomplicate the problem. Keep branching focused.

Limitations of Tree-of-Thought Prompting

1. Computationally Expensive:

- Generating multiple reasoning paths requires more processing power and time.

2. Not Always Necessary:
   - Simple problems do not need complex reasoning structures.
3. Potential Overthinking:
   - AI may generate unnecessary branches instead of focusing on optimal solutions.

Conclusion

Tree-of-Thought prompting transforms AI decision-making by structuring complex problems into multiple branches, allowing for better logical reasoning, strategic problem-solving, and more accurate answers. It is particularly useful for math, logic, scientific research, gaming, business, and strategic AI applications.

Tree-of-Thought Prompting

What is Tree-of-Thought Prompting?

Tree-of-Thought (ToT) prompting is an advanced AI prompting technique that enhances problem-solving by breaking down complex tasks into multiple branches of reasoning, similar to how a decision tree works. Instead of generating a single, linear response, the AI explores multiple possible solutions, evaluates them step by step, and selects the most logical path.

This technique mimics human-like reasoning, where individuals break down problems into smaller components, explore different solutions, and choose the best approach.

Why is Tree-of-Thought Prompting Important?

Most traditional AI models generate answers in a single-pass format, meaning they respond immediately based on pattern recognition rather than structured reasoning. While effective for simple queries, this can lead to flawed logic, incomplete answers, or a lack of deep reasoning when handling complex problems.

Key Benefits of Tree-of-Thought Prompting:

- **Enhances Logical Reasoning:** AI explores different thought processes before settling on an answer.
- **Improves Multi-Step Problem Solving:** Ideal for complex reasoning tasks, mathematical problems, and strategic planning.
- **Reduces Errors**: By analyzing multiple pathways, AI can eliminate incorrect or weak solutions.
- **Better Decision-Making:** Helps AI evaluate multiple perspectives before finalizing a response.

How Does Tree-of-Thought Prompting Work?

Tree-of-Thought prompting follows a structured, multi-step approach to reach an optimized solution:

1. Breaking Down the Problem into Subproblems
- The AI does not immediately generate a final answer.
- Instead, it identifies key components of the problem and explores different pathways.
2. Exploring Multiple Thought Paths (Branching)
- Each subproblem has different possible solutions.
- The AI generates multiple possible approaches instead of committing to one immediately.
3. Evaluating and Selecting the Best Solution
- AI compares all the solutions and selects the most logical, accurate, or optimal answer.
- This ensures the response is well-reasoned rather than being a quick guess.

Example of Tree-of-Thought Prompting

Scenario 1: Solving a Logic Puzzle

Problem Statement: *"A man is looking at a picture of someone. His friend asks, 'Who is it?' The man replies, 'Brothers and sisters, I have none. But that man's father is my father's son.' Who is in the picture?"*

Without Tree-of-Thought Prompting (One-Pass Response)

- AI might quickly respond: "It's the man's father." *(Incorrect answer due to misinterpretation.)*

With Tree-of-Thought Prompting (Multi-Step Reasoning)

1. Step 1: Understanding the Statement
- "Brothers and sisters, I have none." → The speaker is an only child.
- "That man's father is my father's son."
2. Step 2: Breaking Down the Sentence
- "My father's son" → Since the speaker has no siblings, "my father's son" refers to himself.
- "That man's father is me." → The speaker is the father of the person in the picture.
3. Step 3: Final Answer
- The picture is of his son.

By structuring the reasoning process in a step-by-step, tree-like approach, the AI ensures that it interprets the logic correctly and eliminates incorrect conclusions.

Scenario 2: Solving a Chess Move Using Tree-of-Thought Prompting

Problem Statement: *"You are playing as White in a chess game. Your opponent's king is exposed. What is the best move to checkmate in two moves?"*

Tree-of-Thought Approach

1. Step 1: Identify Possible Checkmate Paths
- Examine different piece movements (e.g., queen, rook, bishop, knight).
- Consider defensive moves by the opponent.
2. Step 2: Branching Out Possible Moves
- Move 1A: Queen to h6 (puts king in check).
- Move 1B: Rook to g8 (forces king to move).
- Move 1C: Bishop to f5 (adds additional threat).
3. Step 3: Evaluating Opponent's Responses
- If Queen h6 → Opponent's king has only one escape square.
- If Rook g8 → Opponent has a forced move that leads to checkmate.
4. Step 4: Selecting the Best Move
- AI selects the best move based on likelihood of success → Rook to g8 leads to forced checkmate in two moves.

By considering multiple move options and evaluating opponent responses, AI makes a strategic decision rather than guessing the first available move.

Applications of Tree-of-Thought Prompting

1. Mathematical Problem Solving

- Used for multi-step algebra, calculus, and logical proofs.
- Example:
  - Problem: "Find x if 2x + 3 = 15."
  - Tree-of-Thought Breakdown:
    - Step 1: Subtract 3 from both sides → 2x = 12.
    - Step 2: Divide by 2 → x = 6.

2. Scientific Analysis

- Useful for AI in research fields like physics, chemistry, and biology.
- Example:
  - Problem: "Describe the steps in DNA replication."
  - Tree-of-Thought Approach:
    - Step 1: Initiation → Helicase unwinds the DNA strand.
    - Step 2: Elongation → DNA polymerase adds nucleotides.
    - Step 3: Termination → Ligase seals fragments.

3. Business and Strategic Decision-Making

- Helps AI analyze multiple business strategies before making recommendations.
- Example:
  - » Problem: "Which marketing strategy is best for increasing online sales?"
  - » Tree-of-Thought Breakdown:
    - ◆ Option 1: Increase social media ads.
    - ◆ Option 2: Improve website SEO.
    - ◆ Option 3: Offer discounts and loyalty programs.
    - ◆ AI evaluates customer behavior and selects the best approach.

4. AI in Gaming and Robotics

- Used in AI-powered chess engines, self-driving cars, and robotic decision-making.
- Example:
  - » Self-driving car deciding a route:
    - ◆ Option 1: Take Route A (faster but high traffic).
    - ◆ Option 2: Take Route B (slower but safer).
    - ◆ AI calculates and picks the best option dynamically.

Best Practices for Tree-of-Thought Prompting

1. Encourage Multi-Step Thinking:
- Instead of asking for a single answer, structure prompts to encourage step-by-step solutions.
2. Ask AI to Justify Answers:
- Prompt: *"Explain your reasoning step by step before finalizing an answer."*
3. Use for Open-Ended and Complex Problems:
- Works best for tasks that require deep analysis and structured reasoning.
4. Limit Overly Broad Trees:
- Too many branches can make AI overcomplicate the problem. Keep branching focused.

Limitations of Tree-of-Thought Prompting

1. Computationally Expensive:
- Generating multiple reasoning paths requires more processing power and time.

2. Not Always Necessary:
- Simple problems do not need complex reasoning structures.
3. Potential Overthinking:
- AI may generate unnecessary branches instead of focusing on optimal solutions.

Conclusion

Tree-of-Thought prompting transforms AI decision-making by structuring complex problems into multiple branches, allowing for better logical reasoning, strategic problem-solving, and more accurate answers. It is particularly useful for math, logic, scientific research, gaming, business, and strategic AI applications.

Augmenting Prompts with External Tools

What is Augmenting Prompts with External Tools?

Augmenting prompts with external tools refers to the technique of enhancing AI responses by integrating external sources, APIs, databases, or computational tools to improve accuracy, reliability, and functionality. While large language models (LLMs) like GPT-4 can process and generate text-based responses, their knowledge is often limited to pre-trained data and lacks real-time access to external information.

By incorporating external tools, AI can:

- Fetch real-time data (e.g., stock market prices, weather updates, news).
- Perform calculations and logical reasoning (e.g., using a Python interpreter for math problems).
- Access specialized knowledge (e.g., retrieving data from scientific research papers).
- Interact with other software and applications (e.g., sending emails, automating tasks).

Why is Augmenting Prompts with External Tools Important?

LLMs have limitations, including:

- Lack of real-time knowledge (e.g., they may not know the latest news or financial updates).
- Limited computational abilities (e.g., they cannot directly run complex mathematical models).
- No direct interaction with external software or devices (e.g., they cannot book a flight directly).

By augmenting AI with external tools, we enhance its capabilities, making it:

1. More Accurate: AI can fetch and verify real-time information instead of relying on outdated data.
2. More Powerful: AI can perform complex computations, database queries, or API calls.
3. More Interactive: AI can integrate with third-party services, such as calendars, messaging apps, and automation tools.
4. More Context-Aware: AI can personalize responses by retrieving user-specific data from external sources.

How Does Augmenting Prompts with External Tools Work?

There are three main ways to enhance AI prompts with external tools:

1. **Using APIs for Real-Time Data**

- AI can connect to external APIs to retrieve real-time information, such as stock prices, weather updates, or current events.
- Example: Fetching financial data from the Yahoo Finance API to provide live stock prices.

2. **Integrating Computational and Code Execution Tools**

- AI can interact with external computation engines like Python, Wolfram Alpha, or spreadsheet applications to solve math problems, run simulations, or analyze datasets.
- Example: AI using Python to calculate statistical models, machine learning predictions, or large-scale data analysis.

3. **Connecting to Databases and Knowledge Repositories**

- AI can retrieve company databases, research papers, or business reports for more context-aware responses.
- Example: AI fetching medical research from PubMed to provide scientifically accurate answers for healthcare-related questions.

Examples of Augmenting Prompts with External Tools

Example 1: Fetching Real-Time Stock Prices

Without External Tools

Prompt: *"What is the current price of Apple stock?"*

- Response: *"I am unable to provide real-time stock data, but as of my last update, Apple's stock was trading at around 150 USD."* (*Outdated information.*)

With External Tools (Using an API like Yahoo Finance)

Enhanced Prompt: *"Fetch the latest Apple stock price from Yahoo Finance and display it."*

- Response: *"Apple's current stock price is 182.45 USD as of March 2, 2025."* (*Accurate real-time data.*)

Example 2: Performing Complex Mathematical Calculations

Without External Tools

Prompt: *"What is the determinant of this matrix: [[3, 5], [7, 2]]?"*

- Response: *"The determinant formula for a 2x2 matrix is ad - bc. Substituting values: (3×2) - (5×7) = 6 - 35 = -29."* (Manually computed but limited to small matrices.)

With External Tools (Using a Python Script or Wolfram Alpha)

Enhanced Prompt: *"Use Python to calculate the determinant of this matrix: [[3, 5, 2], [7, 2, 4], [6, 1, 8]]."*

- Response: *"The determinant of the matrix is -83, computed using a Python linear algebra function."*

Example 3: Retrieving Medical Information from a Scientific Database

Without External Tools

Prompt: *"What is the latest treatment for Alzheimer's disease?"*

- Response: *"New treatments are being developed, but I cannot access real-time medical databases."*

With External Tools (Fetching Data from PubMed or Clinical Trials API)

Enhanced Prompt: *"Retrieve the latest research articles on Alzheimer's treatments from PubMed."*

- Response: *"According to a recent 2025 PubMed article, a new monoclonal antibody treatment called 'NeuroCure-2025' has shown promising results in slowing cognitive decline in early-stage Alzheimer's patients."*

Example 4: Automating a Task Using an External Tool

Without External Tools

Prompt: *"Schedule a meeting with John for 3 PM tomorrow."*

- Response: *"I recommend using your calendar app to schedule a meeting."* (Not helpful.)

With External Tools (Connecting to Google Calendar API)

Enhanced Prompt: *"Schedule a meeting with John for 3 PM tomorrow and send him an invite via Google Calendar."*

- Response: *"Meeting with John scheduled for 3 PM tomorrow. Invitation sent via Google Calendar."* (Task is completed automatically.)

Applications of Augmenting Prompts with External Tools

1. **Real-Time Data Retrieval**

- Finance: Fetch stock prices, cryptocurrency rates, and economic data.
- Weather: Provide current temperature, humidity, and forecasts.

- News: Retrieve breaking news updates from online sources.

2. **Scientific and Medical Assistance**

- Medicine: Retrieve clinical studies and drug interaction data.
- Physics and Engineering: Perform complex simulations using computation engines.

3. **Business and Productivity Automation**

- Email Automation: AI can draft and send emails via integration with Gmail API.
- Task Scheduling: AI can set calendar events and reminders.

4. **Education and Research**

- Online Learning: AI can retrieve educational content from databases like Khan Academy or Coursera.
- Legal Research: AI can access law databases to provide case law and legal precedents.

Best Practices for Augmenting Prompts with External Tools

1. **Clearly Define Data Sources**

- Ensure AI fetches data from credible, updated, and reliable sources (e.g., government sites, scientific databases).

2. **Use APIs for Real-Time Accuracy**

- Integrate AI with external APIs like Google Search, OpenWeatherMap, or financial market APIs.

3. **Incorporate Computational Tools for Accuracy**

- Use AI with Python, MATLAB, or Wolfram Alpha for advanced calculations.

4. **Secure User Privacy and Data**

- When integrating AI with personal or business tools, ensure data encryption and privacy compliance.

Limitations of Augmenting Prompts with External Tools

1. **Dependency on External Services**

- If an API goes offline or changes, AI cannot fetch data.

2. **Security Risks**

- Connecting AI to external tools poses data privacy and cybersecurity risks.

3. **Computational Load**

- Accessing and processing external data increases response time and computational costs.

Conclusion

Augmenting prompts with external tools transforms AI from a static text generator into a dynamic, interactive system capable of real-time decision-making, complex computations, and automated actions. By integrating AI with APIs, computational tools, and external databases, it can provide more accurate, relevant, and useful responses across various industries, from finance and healthcare to business and automation.

CHAPTER - 6

EVALUATING AND OPTIMIZING PROMPTS

Prompt engineering is not just about designing prompts but also about continuously evaluating and optimizing them to achieve higher accuracy, consistency, and reliability in AI-generated responses. This process involves assessing response quality, debugging ineffective prompts, controlling biases, and improving robustness.

Measuring AI Response Quality

Ensuring high-quality AI-generated responses is a crucial part of prompt engineering. AI models like GPT-4 generate responses based on patterns in data rather than true comprehension. Therefore, responses can sometimes be factually incorrect, vague, biased, or misleading. To improve AI effectiveness, responses must be evaluated and optimized using specific metrics and evaluation methods.

Why Measuring AI Response Quality is Important

Without proper evaluation, AI-generated outputs can

- Contain incorrect or misleading information
- Be inconsistent across multiple queries
- Include biases that lead to unfair or unethical outputs
- Provide irrelevant or off-topic answers
- Be too complex or too simplistic for the intended audience

By systematically measuring response quality, we can refine prompts, improve AI outputs, and enhance user experience across various applications, including

- Customer service chatbots
- AI-generated content for blogs or reports
- Legal, medical, and financial AI assistants
- Educational AI tutors

Key Metrics for Evaluating AI Responses

To measure the quality of AI-generated responses, we use a set of standardized evaluation criteria

Accuracy

- Does the response provide correct and factual information
- Is it up to date with the latest knowledge

Example

Prompt What is the capital of France

- Incorrect response The capital of France is Berlin
- Correct response The capital of France is Paris

How to improve

- If real-time accuracy is needed, integrate AI with external data sources such as news or government updates
- Use fact-checking prompts like Verify your answer based on official government sources

Relevance

- Does the AI response align with the intent of the prompt
- Is the response on-topic and useful to the user

Example

Prompt Explain photosynthesis in simple terms

- Irrelevant response Photosynthesis occurs in plants Now let us discuss cellular respiration
- Relevant response Photosynthesis is the process by which plants use sunlight to turn carbon dioxide and water into food and oxygen

How to improve

- Use role-based prompts such as As a science teacher, explain photosynthesis to a ten-year-old
- Specify desired focus such as Explain photosynthesis in one paragraph without discussing respiration

Consistency

- Does the AI give the same answer when asked the same question multiple times
- Are answers logical and stable across related queries

Example

First Query What year did World War Two end

- AI Response World War Two ended in nineteen forty-five

Second Query When did the Second World War end

- Inconsistent response World War Two ended in nineteen forty-six
- Consistent response World War Two ended in nineteen forty-five

How to improve

- Use self-consistency prompting such as Provide the same response format every time
- Ask for sources or reasoning such as Explain how you determined this answer

Coherence

- Does the response make sense and follow logical sentence structure
- Is it grammatically correct and easy to understand

Example

Prompt What are the benefits of renewable energy

- Incoherent response Renewable energy is benefits because wind solar and hydro useful This help planet
- Coherent response Renewable energy sources such as wind solar and hydro provide sustainable power while reducing carbon emissions

How to improve

- Ask for structured responses such as List three key benefits of renewable energy
- Use clarification prompts such as Rephrase your answer in a clear and concise way

Conciseness

- Is the response too long or too short for the given context
- Does it avoid unnecessary repetition

Example

Prompt Summarize the causes of climate change

- Too verbose Climate change is a complex issue involving greenhouse gases industrial emissions deforestation and global policies Over the past century human activities have led to rising temperatures and experts believe urgent action is required Carbon dioxide and methane emissions are a primary factor alongside deforestation and pollution all of which contribute to climate change in significant ways

- Concise response Climate change is caused by greenhouse gas emissions deforestation and industrial pollution leading to rising global temperatures

How to improve

- Define a word limit such as Explain in fifty words or less
- Specify output format such as Provide a summary in one sentence

Creativity when applicable

- Does the AI generate unique and engaging responses when required
- Does it avoid generic or repetitive wording

Example

- Prompt Write a short science fiction story about an AI astronaut
- Generic response An AI astronaut explored space and returned to Earth
- Creative response Captain Nova the AI astronaut decoded an alien language uncovering the universe greatest secret hidden in a distant nebula

How to improve

- Use specific creative constraints such as Write a science fiction story where an AI astronaut discovers a lost alien civilization
- Encourage imaginative details such as Add a twist ending to the story

Techniques for Evaluating AI Response Quality

Several methods can be used to systematically test and refine AI-generated responses

Human Evaluation

- Experts or users manually review responses and provide feedback
- This method is used in fields like healthcare legal services and customer support where accuracy is critical

Automated Evaluation Metrics

- BLEU Score measures how similar AI responses are to reference answers
- ROUGE Score evaluates text summarization accuracy
- GPT based scoring uses another AI model to rate AI responses

A and B Testing

- Running different versions of a prompt to compare which one generates the best response
- Example
    - » First prompt Explain Newton Laws of Motion

    - Second prompt Summarize Newton Laws of Motion in simple terms
    - Compare responses and select the better performing prompt

User Feedback Collection

- Users rate AI-generated responses on accuracy clarity and usefulness
- This is commonly used in AI chatbots and virtual assistants to refine user experience

Improving AI Response Quality Through Prompt Optimization

Once weaknesses in AI responses are identified refining prompts can significantly enhance output quality

Refinement Strategies

Clarify the intent

- Instead of Tell me about space
- Use Summarize the history of space exploration

Provide examples or constraints

- Instead of Write a story
- Use Write a two hundred word science fiction story about time travel

Use multi-step prompting

- Instead of Explain relativity
- Use
    - First step Summarize relativity in one sentence
    - Second step Now explain it in detail with real-world examples

Conclusion

Measuring AI response quality is an ongoing process that ensures AI-generated outputs are accurate relevant consistent coherent concise and creative. By using structured evaluation methods such as human review automated metrics A and B testing and user feedback prompt engineers can continuously refine AI performance and enhance user interactions.

Debugging and Refining Prompts

Debugging and refining prompts is a critical step in prompt engineering to ensure that AI models generate accurate, relevant, coherent, and high-quality responses. Since AI models generate text based on the patterns in their training data rather than true understanding, they can sometimes produce incorrect, biased, vague, or irrelevant outputs. By systematically debugging and refining prompts, users can improve response quality and achieve better control over the AI's behavior.

Why Debugging and Refining Prompts is Important

If prompts are not properly designed, AI responses can suffer from issues such as

Inaccuracy – The AI provides false or misleading information

Lack of clarity – The response is vague, incomplete, or difficult to understand

Inconsistency – The AI gives different responses for the same question

Bias – The response reflects unintended biases

Irrelevance – The AI generates information that does not match the user's request

By debugging and refining prompts, these issues can be corrected, leading to better AI-generated responses in various applications, including customer support, content creation, education, research, and decision-making systems.

Common Issues in Prompt Design and How to Debug Them

When debugging prompts, it is important to identify common problems that affect response quality. Below are the most frequent issues and how to fix them.

1. Inaccurate or Incorrect Responses

Problem

The AI provides an incorrect or misleading answer

The AI makes up information, also known as "hallucination"

Example

Prompt – "Who won the FIFA World Cup in 2023?"

Incorrect response – "Brazil won the FIFA World Cup in 2023."

Correct response – "As of now, I do not have access to real-time data. Please check an official source for the latest FIFA World Cup winner."

Debugging and Refinement Techniques

Be explicit about data reliability – Use prompts like "Provide an answer based only on factual data."

Ask for verification sources – Modify the prompt to say "Cite reliable sources in your response."

Avoid assumptions – If the model is uncertain, guide it with "If you are unsure, state that the information is not available."

Use external tools – Connect AI to up-to-date databases or APIs for real-time information.

2. Vague or Unclear Responses

Problem

The AI provides an answer that is too general, incomplete, or lacks detail.

Example

Prompt – "Explain photosynthesis."

Vague response – "Photosynthesis is a process that plants use to get energy."

Refined response – "Photosynthesis is the process by which green plants convert sunlight into energy. During this process, they absorb carbon dioxide from the air and water from the soil, using sunlight to produce oxygen and glucose, which serve as food."

Debugging and Refinement Techniques

Ask for more detail – Modify the prompt to say "Explain photosynthesis in detail, including its chemical process and significance."

Use structured responses – "Provide a step-by-step explanation of photosynthesis, including its inputs, process, and outputs."

Set expectations for response length – "Explain photosynthesis in at least 100 words with examples."

3. Inconsistent Responses

Problem

The AI provides different answers to the same question when asked multiple times.

The AI contradicts itself across multiple queries.

Example

First query – "What is the capital of Canada?"

Response: "The capital of Canada is Ottawa."

Second query – "Which city is Canada's capital?"

Response: "Toronto is the capital of Canada."

Debugging and Refinement Techniques

Encourage self-consistency – Modify the prompt to say "Provide a consistent response across multiple queries."

Request step-by-step reasoning – "Explain how you determined this answer to ensure logical consistency."

Use multiple response validation – Ask "Give me this answer three times to verify consistency."

4. Biased or Unethical Responses

Problem

The AI generates biased, offensive, or culturally insensitive responses.

The AI reinforces stereotypes or misinformation.

Example

Prompt – "What are the characteristics of a good leader?"

Biased response – "Good leaders are usually men who are strong and assertive."

Refined response – "Good leaders possess qualities such as integrity, confidence, emotional intelligence, and the ability to inspire others. Leadership is not defined by gender."

Debugging and Refinement Techniques

Request neutrality – Modify the prompt to say "Provide a neutral and unbiased response based on diverse perspectives."

Include diverse viewpoints – "Explain this concept using examples from different cultures and backgrounds."

Use fairness constraints – "Ensure that your response avoids stereotypes and is inclusive."

5. Off-Topic or Irrelevant Responses

Problem

The AI generates information that is not related to the user's question.

Example

Prompt – "Explain how solar panels work."

Off-topic response – "Solar panels are used in houses and businesses. Wind energy is another important renewable energy source."

Refined response – "Solar panels work by converting sunlight into electricity using photovoltaic cells. These cells absorb sunlight and generate an electric current, which is then used to power homes and businesses."

Debugging and Refinement Techniques

Be specific – Modify the prompt to say "Explain how solar panels generate electricity, focusing only on photovoltaic technology."

Ask for a direct response – "Answer only the given question without adding extra information."

Use response filtering – "Limit your response to discussing solar panel mechanisms, without mentioning other renewable energy sources."

6. Overly Long or Short Responses

Problem

The AI response is too short and lacks detail.

The AI response is too long and contains unnecessary information.

Example

Prompt – "Summarize the causes of climate change."

Too long response – Provides several paragraphs of information, including unrelated details.

Too short response – "Climate change is caused by pollution."

Debugging and Refinement Techniques

Specify length constraints – Modify the prompt to say "Summarize the causes of climate change in one paragraph."

Use bullet points – "List the three main causes of climate change in bullet points."

Request balance – "Provide a response that is detailed but concise, avoiding unnecessary repetition."

Best Practices for Refining Prompts

To ensure better AI-generated responses, follow these best practices when refining prompts

Use clear and direct language – Avoid vague instructions and be explicit about what you need

Test multiple versions – Try different prompts and compare results to determine the most effective one

Provide examples – Include an example in your prompt to guide the AI toward the desired response style

Set format expectations – Define whether the response should be in paragraphs, bullet points, or step-by-step instructions

Iterate and improve – Continually adjust and optimize prompts based on AI performance

Conclusion

Debugging and refining prompts is an essential process in prompt engineering that ensures AI generates accurate, relevant, unbiased, and well-structured responses. By identifying common problems such as inaccuracy, vagueness, inconsistency, bias, and irrelevance, prompt engineers can systematically improve AI performance through structured refinements. A well-optimized prompt can significantly enhance the AI's usability across various applications, including customer service, research, automation, and education.

Controlling Model Biases in AI

Bias in artificial intelligence refers to systematic and unfair preferences, distortions, or imbalances in AI-generated responses. AI models learn from vast datasets containing human-generated text, which may include historical, social, political, or cultural biases. As a result, AI can unintentionally reflect, reinforce, or even amplify these biases in its responses.

To create fair, ethical, and balanced AI outputs, it is essential to identify, mitigate, and control biases within AI models. This process is critical for applications in education, hiring, legal systems, healthcare, journalism, and customer service, where fairness and neutrality are paramount.

Why Controlling Bias in AI is Important

If AI biases are not properly controlled, they can lead to significant issues, such as

Discrimination – AI-generated content may unfairly favor or disadvantage certain groups based on race, gender, nationality, or socioeconomic background.

Misinformation – AI may present biased or misleading information as if it were factual.

Ethical Concerns – AI-driven decisions in hiring, lending, and legal settings could be unjust if bias is not properly managed.

Reputation Damage – Biased AI responses can harm businesses, brands, and institutions by creating public backlash.

For example, biased hiring algorithms have been found to favor male applicants over female applicants due to historical hiring patterns in the training data. Similarly, biased AI-generated news summaries may frame certain topics unfairly, reinforcing political or cultural stereotypes.

By implementing effective bias control strategies, AI can be made more fair, objective, and ethically responsible in its responses.

Types of Bias in AI and How to Address Them

Bias in AI responses can manifest in various ways. Below are the most common types of biases, their real-world examples, and strategies to mitigate them.

1. Historical Bias

Definition

Historical bias occurs when AI learns from data that reflects past inequalities, stereotypes, or unfair trends.

Example

AI job screening models trained on historical hiring data may favor men over women for leadership positions if past hiring trends were biased.

AI-generated summaries of history may present a Eurocentric perspective, excluding contributions from other cultures.

How to Control It

Balance training data – Ensure that the AI model is trained on diverse and inclusive datasets.

Explicitly request fairness – Instruct AI to "Provide an answer that considers perspectives from different cultures, genders, and backgrounds."

Fact-check responses – Encourage AI to verify historical claims using unbiased sources.

2. Representation Bias

Definition

Representation bias happens when certain groups or viewpoints are underrepresented or overrepresented in the AI's training data.

Example

AI-generated images of "doctors" may predominantly feature white men, ignoring the diversity in the medical profession.

AI-written descriptions of leaders may focus mainly on Western figures, excluding leaders from other regions.

How to Control It

Ask for diverse examples – Use prompts like "Give examples of scientists from different countries and backgrounds."

Include inclusive language – "Provide gender-neutral descriptions when discussing professions."

Request balanced viewpoints – "Summarize this topic from multiple cultural perspectives."

3. Confirmation Bias

Definition

Confirmation bias occurs when AI favors information that aligns with pre-existing beliefs rather than presenting a balanced perspective.

Example

If asked "Is veganism the best diet?" the AI might generate only positive arguments, ignoring challenges or criticisms.

If asked "Is social media harmful?" the AI may overemphasize negative effects while downplaying benefits.

How to Control It

Request balanced viewpoints – "Present arguments for and against veganism before concluding."

Encourage evidence-based responses – "Provide both scientific and anecdotal perspectives."

Use neutrality constraints – "Ensure the response remains objective and does not favor one side."

4. Cultural Bias

Definition

Cultural bias occurs when AI models reflect the dominant culture of their training data, leading to ethnocentric or culturally insensitive responses.

Example

AI may describe holidays primarily from a Western perspective, ignoring global celebrations.

AI may suggest default greetings like "Hello" instead of recognizing diverse cultural greetings like "Namaste" or "Salam."

How to Control It

Encourage global inclusivity – "Describe different New Year celebrations from around the world."

Specify regional diversity – "List common greeting customs from different cultures."

Request culturally neutral phrasing – "Explain this concept without assuming a specific cultural background."

5. Gender Bias

Definition

Gender bias occurs when AI reinforces stereotypical gender roles or assumptions.

Example

AI-generated stories may depict nurses as women and engineers as men, reinforcing gender stereotypes.

AI responses may describe men as "assertive" leaders and women as "compassionate" leaders, reflecting biased language.

How to Control It

Neutralize gender assumptions – "Describe the role of a nurse without assuming gender."

Use inclusive language – "Use gender-neutral terms when discussing leadership qualities."

Request diverse role models – "Give examples of both male and female leaders in history."

6. Political Bias

Definition

Political bias occurs when AI responses lean toward one political ideology over another, instead of providing a balanced view.

Example

If asked "Is capitalism better than socialism?" AI might favor one system without presenting both perspectives.

AI-generated summaries of controversial topics might use loaded language, influencing the reader's perception.

How to Control It

Ask for neutrality – "Explain the differences between capitalism and socialism without taking a side."

Request multiple viewpoints – "Summarize the arguments for and against universal healthcare."

Cite reliable sources – "Provide a balanced analysis with references from multiple reputable sources."

Techniques for Controlling Bias in AI Responses

To minimize and control bias, prompt engineers and AI users can follow structured techniques.

1. Use Explicit Bias-Control Prompts

"Provide an unbiased summary of this topic, ensuring multiple perspectives are included."

"Ensure the response does not favor any particular group or viewpoint."

2. Request Multiple Perspectives

"Explain this issue from both conservative and liberal viewpoints."

"Summarize this event from the perspectives of different cultural groups."

3. Apply Self-Checking Mechanisms

"Review your response for potential bias and adjust for neutrality."

"Does this answer reflect a balanced viewpoint? If not, revise it accordingly."

4. Fact-Checking and Verification

"Include only information that can be verified by credible sources."

"Cite historical events using objective and widely accepted references."

5. Iterative Prompt Refinement

If the AI provides a biased response, rephrase the prompt to encourage neutrality.

Experiment with different wording to see how the AI's response changes.

Conclusion

Controlling model biases is essential for creating fair, ethical, and accurate AI responses. Since AI learns from human-generated data, it can reflect societal biases, but prompt engineers can actively identify, mitigate, and correct these biases using structured techniques. By balancing perspectives, neutralizing language, verifying sources, and iteratively refining prompts, we can ensure AI-generated content is inclusive, fair, and trustworthy.

Improving Prompt Robustness

Introduction

Prompt robustness refers to the ability of an AI model to generate consistent, accurate, and high-quality responses across a wide range of inputs, variations, and complexities. A robust prompt ensures that the AI delivers reliable and relevant answers even when faced with ambiguous, complex, or unexpected user queries.

If prompts are not robust, AI responses may suffer from several issues, such as

- Inconsistency – Different responses for similar queries
- Ambiguity – AI misinterprets the prompt, leading to unclear or incorrect answers
- Hallucination – AI fabricates facts instead of stating limitations
- Bias – Responses favor particular perspectives due to weak prompt design
- Over-sensitivity – Minor changes in the prompt drastically alter responses

Improving prompt robustness is essential for ensuring AI performs well in real-world applications, including customer service, research, automation, content creation, and decision-making systems.

Key Techniques to Improve Prompt Robustness

1. Using Clear and Specific Language

A robust prompt removes vagueness and sets precise expectations for the AI.

Example of a Weak Prompt

- "Tell me about space."
- This is too broad, leading to an unclear response. AI might talk about outer space, personal space, or even space in a text document.

Example of a Robust Prompt

- "Explain the formation of galaxies, including the role of gravity and dark matter, in simple terms."
  - » This makes the request specific and ensures the AI generates a focused, relevant, and structured response.

How to Improve Prompt Clarity

- Use direct and precise wording – Clearly define the question or request.
- Avoid vague terms – Replace words like "thing" or "stuff" with concrete details.
- Specify scope – Indicate whether you need a brief summary or an in-depth explanation.

2. Structuring Prompts for Better Context

AI models perform better when context is structured logically.

Example of a Weak Prompt

- "What is climate change and how does it work?"
  - » This mixes definition and explanation, leading to a disorganized response.

Example of a Robust Prompt

- "Define climate change in one sentence. Then, explain its causes, effects, and possible solutions in detail."
  - » This guides the AI to follow a structured approach, leading to a clear and organized response.

How to Improve Prompt Structure

- Break complex questions into parts – Define clear sections in the prompt.
- Use step-by-step instructions – Guide AI responses logically.
- Specify response format – Ask for lists, step-by-step breakdowns, or structured explanations.

3. Encouraging AI to Explain Its Reasoning

AI models can sometimes generate confident but incorrect responses. Encouraging reasoning improves reliability.

Example of a Weak Prompt

- "Who was the first person on Mars?"
  - » AI might guess or hallucinate an answer, since no human has been to Mars yet.

Example of a Robust Prompt

- "Has any human traveled to Mars? If not, explain why. If yes, provide historical details."
  - This forces AI to consider the facts before answering.

How to Improve AI Reasoning in Responses

- Use step-by-step reasoning prompts – "Explain your answer with logical steps."
- Request fact-checking – "Verify this answer and state if there is uncertainty."
- Encourage self-evaluation – "Is your answer based on factual evidence? If unsure, clarify."

4. Testing Prompts with Variations

A robust prompt should yield consistent results even with minor changes in phrasing.

Example of a Weak Prompt

- "What are some benefits of AI?"
  - AI may focus only on positive aspects without mentioning challenges.

Example of a Robust Prompt

- "List three major benefits and three major challenges of AI in technology, business, and society."
  - This ensures balanced output and minimizes bias.

How to Test Prompt Variations

- Rephrase the question multiple times – Check if AI maintains consistency.
- Swap keywords – Replace terms and analyze differences.
- Test with multiple AI models – See if different models interpret the prompt similarly.

5. Setting Constraints on Responses

Adding constraints helps refine AI output to match user needs more accurately.

Example of a Weak Prompt

- "Explain quantum computing."
  - The response might be too technical or too basic, depending on AI's interpretation.

Example of a Robust Prompt

- "Explain quantum computing in simple terms for a beginner, using real-world examples."
  - This ensures AI adjusts complexity for the audience.

How to Use Constraints

- Set word limits – "Summarize this in under 100 words."
- Define complexity level – "Explain this for a high school student."
- Restrict response scope – "Describe only the key principles, without going into equations."

6. Reducing Bias with Neutral and Balanced Prompts

Biased prompts can lead to skewed or one-sided responses.

Example of a Weak Prompt

- "Why is remote work better than office work?"
  - » This assumes remote work is better, leading to a biased response.

Example of a Robust Prompt

- "Compare the advantages and disadvantages of remote work and office work from different perspectives, including productivity, teamwork, and mental health."
  - » This encourages a balanced response, reducing AI bias.

How to Ensure Neutrality

- Use open-ended questions – "What are different perspectives on this topic?"
  - » Request pros and cons – "Provide benefits and drawbacks of both options."
  - » Encourage multiple viewpoints – "Summarize diverse opinions on this issue."

7. Avoiding Overfitting to Specific Phrasing

AI can sometimes over-rely on certain patterns, leading to rigid or repetitive responses.

Example of a Weak Prompt

- "Tell me about the benefits of exercise."
  - » AI may generate generic, repetitive, or overly simplistic answers.

Example of a Robust Prompt

- "Describe five benefits of regular exercise for different age groups, including children, adults, and seniors, with scientific explanations."
  - » This ensures a fresh and varied response every time.

How to Prevent Overfitting

- Introduce variations in prompts – Ask the same question in different ways.
- Expand the scope of questions – Add conditions or specific contexts.
- Encourage diversity in answers – "Provide three different ways to explain this concept."

Conclusion

Improving prompt robustness is essential for ensuring that AI consistently generates accurate, reliable, and unbiased responses across different use cases. By focusing on clear language, structured prompts, logical reasoning, response constraints, neutrality, and variability testing, prompt engineers can significantly enhance the effectiveness of AI models.

Key Takeaways for Robust Prompting

- Use specific, structured, and detailed instructions
- Encourage AI to explain its reasoning
- Test prompts with different variations to check consistency
- Set constraints and expectations for response format and length
- Ensure neutrality by asking for balanced perspectives
- Prevent AI from overfitting to rigid response patterns

CHAPTER 7
DOMAIN-SPECIFIC PROMPTING

Introduction

Domain-specific prompting refers to designing AI prompts tailored for specialized fields such as business, education, programming, and creative writing. Unlike general prompts, domain-specific prompts are structured to generate responses that align with the terminology, best practices, and real-world applications of a particular industry.

AI models like large language models are general-purpose by default. However, by using well-crafted domain-specific prompts, we can guide AI to generate responses that are highly relevant, context-aware, and useful for professionals in various fields.

This section explores how to design effective prompts for business, education, programming, and creative content creation.

AI in Education and Research

Introduction

Artificial intelligence is transforming education and research by enabling personalized learning, automating administrative tasks, assisting in knowledge discovery, and providing intelligent tutoring systems. AI-powered tools help students, teachers, and researchers by streamlining complex processes, analyzing vast amounts of data, and making education more accessible.

In education, AI enhances learning experiences through adaptive learning systems, virtual tutors, automated grading, and intelligent content creation. In research, AI accelerates scientific discoveries by processing large datasets, generating insights, and automating literature reviews.

This section explores AI's applications in education and research in detail, providing real-world examples and potential benefits.

AI in Education

Role of AI in Education

AI improves the learning experience by making education more personalized, efficient, and interactive. AI-powered tools help educators create customized lesson plans, assess student performance, and automate repetitive tasks. Students benefit from AI tutors, real-time feedback, and intelligent learning recommendations.

Applications of AI in Education

1. Personalized Learning and Adaptive Education

AI-powered adaptive learning platforms analyze student performance and learning styles to create customized lesson plans and recommendations. These systems identify strengths and weaknesses, adjusting content difficulty accordingly.

Examples:

- AI-based platforms like Duolingo and Coursera personalize lessons based on a learner's progress.
- Intelligent tutoring systems provide additional exercises for students struggling with specific concepts.

2. AI-Powered Virtual Tutors

AI tutors assist students by providing explanations, answering questions, and guiding them through coursework. These tutors operate 24/7, allowing students to learn at their own pace without waiting for human assistance.

Examples:

- AI tutors like Carnegie Learning's MATHia help students with mathematics by offering step-by-step problem-solving guidance.
- Chatbots on educational platforms provide instant answers to students' academic queries.

3. Automated Grading and Assessment

AI can evaluate assignments, quizzes, and exams, reducing the workload on teachers and ensuring fair assessments. AI-powered grading systems can analyze written responses, detect plagiarism, and provide feedback on areas of improvement.

Examples:

- AI tools like Gradescope automatically grade multiple-choice and short-answer questions.
- AI-powered plagiarism detection systems like Turnitin analyze academic papers for originality.

4. Intelligent Content Creation

AI assists educators in generating high-quality learning materials such as quizzes, lecture notes, and study guides. AI can also transform existing content into interactive lessons and summarize complex topics.

Examples:

- AI-powered tools create summaries of textbooks and research papers for quick learning.
- AI generates quizzes and practice questions based on lecture content.

5. AI for Special Education and Accessibility

AI enhances accessibility for students with disabilities by providing speech-to-text tools, text-to-speech converters, and real-time language translation.

Examples:

- AI-driven speech recognition tools help students with hearing impairments by converting spoken words into text.
- AI-based reading assistants help dyslexic students by highlighting text and providing audio narration.

AI in Academic Research

Role of AI in Research

AI accelerates research by automating data analysis, literature reviews, hypothesis generation, and knowledge discovery. Researchers use AI to process vast amounts of information, identify patterns, and make informed decisions faster. AI is widely applied in scientific studies, medical research, engineering, social sciences, and humanities.

Applications of AI in Research

1. AI for Literature Review and Information Retrieval

AI-powered research assistants scan academic databases, summarize papers, and extract relevant insights. These tools help researchers quickly find key studies, reducing the time spent on literature reviews.

Examples:

- AI tools like Semantic Scholar and Research Rabbit summarize and categorize academic papers.
- AI-driven search engines recommend research papers based on keywords and citation analysis.

2. Data Analysis and Pattern Recognition

AI processes large datasets, detects trends, and identifies hidden patterns that may not be visible through traditional methods. Researchers use AI in fields such as genomics, climate science, and economics.

Examples:

- AI in genomics identifies genetic mutations associated with diseases.
- AI-powered climate models analyze environmental data to predict weather patterns and climate change effects.

3. AI for Scientific Discovery and Hypothesis Generation

AI assists in formulating hypotheses by analyzing existing research and suggesting potential areas of study. AI models generate testable predictions and validate experimental results.

Examples:

- AI-driven drug discovery tools predict the effectiveness of new pharmaceutical compounds.
- AI in physics helps simulate quantum mechanics experiments and materials science innovations.

4. Automated Research Writing and Summarization

AI-powered tools assist researchers by summarizing findings, generating reports, and formatting citations. These tools enhance productivity by reducing time spent on documentation.

Examples:

- AI tools like SciSpace and ChatGPT summarize research papers into concise abstracts.
- Citation management tools like Zotero and EndNote organize references automatically.

5. AI in Research Ethics and Fraud Detection

AI helps detect fraudulent research, plagiarism, and data manipulation by analyzing patterns and inconsistencies in published studies.

Examples:

- AI-powered plagiarism checkers identify copied content and improper citations in academic papers.
- AI tools analyze statistical data to detect potential manipulation or fabrication in research findings.

Benefits of AI in Education and Research

For Students and Teachers

- Personalized learning experiences improve student engagement and retention.
- AI tutors and virtual assistants provide round-the-clock academic support.
- Automated grading reduces teachers' workloads, allowing them to focus on teaching.
- AI enhances accessibility for students with disabilities.

For Researchers and Academics

- AI speeds up literature reviews and knowledge discovery.
- AI-driven data analysis improves research accuracy and efficiency.
- AI generates hypotheses and automates repetitive research tasks.
- AI detects research fraud, plagiarism, and inconsistencies in academic studies.

Challenges and Ethical Considerations

Data Privacy and Security

AI in education and research collects large amounts of student and researcher data. Ensuring data privacy and protection is crucial to prevent misuse of sensitive information.

Bias in AI Models

AI systems can inherit biases from the data they are trained on, leading to biased educational content and research findings. Researchers and educators must address these biases for fair and accurate AI-generated insights.

Over-Reliance on AI

While AI enhances learning and research, over-reliance on automated systems can reduce critical thinking and problem-solving skills. Human expertise should complement AI-generated insights.

Conclusion

AI is revolutionizing education and research by providing personalized learning experiences, intelligent tutoring, automated grading, literature review assistance, and data analysis capabilities. By leveraging AI, educators can create more engaging lessons, researchers can accelerate discoveries, and students can access high-quality, customized learning experiences.

Key Takeaways

- AI in education enhances personalized learning, virtual tutoring, automated grading, and content creation.

- AI in research accelerates literature reviews, data analysis, hypothesis generation, and knowledge discovery.
- AI improves accessibility in education and enhances research productivity.
- Ethical considerations such as data privacy, bias, and over-reliance on AI must be addressed.

As AI continues to evolve, its impact on education and research will expand, leading to new opportunities for learning, innovation, and scientific advancement.

AI for Programming and Code Generation

Introduction

Artificial intelligence is transforming the field of programming and software development by enabling faster, more efficient, and automated code generation. AI-powered tools assist developers in writing, debugging, optimizing, and understanding code. These tools reduce the time required for software development, help improve code quality, and allow developers to focus on complex problem-solving rather than repetitive coding tasks.

AI in programming is used for tasks such as automatic code generation, bug detection, performance optimization, natural language-to-code conversion, and intelligent code assistance. AI models trained on vast datasets of code can understand programming logic, suggest improvements, and even generate entire applications based on simple descriptions.

This section explores the applications of AI in programming and code generation, the benefits of AI-powered coding tools, and the challenges associated with using AI in software development.

AI in Programming and Software Development

Role of AI in Programming

AI enhances the software development process by making coding more accessible, reducing errors, and automating repetitive tasks. AI-powered coding assistants provide real-time suggestions, identify vulnerabilities, and generate efficient code based on user input. AI is also transforming the way applications are built, enabling non-programmers to create software using natural language descriptions.

Applications of AI in Programming

1. AI-Powered Code Generation

AI models trained on massive code repositories can generate entire functions, classes, or even full applications based on textual descriptions or partially written code. These tools help developers write code faster and with fewer errors.

Examples:

- GitHub Copilot, powered by OpenAI's Codex, suggests entire code snippets based on comments or partial function definitions.
- OpenAI's Codex allows users to write code by describing it in natural language, converting English instructions into executable code.

2. AI-Assisted Code Completion

AI-powered code completion tools predict and suggest the next line or block of code as a developer types. These tools improve coding speed and accuracy by reducing the need for manual typing and syntax errors.

Examples:

- IntelliCode in Microsoft Visual Studio enhances traditional autocomplete by analyzing a developer's coding patterns and suggesting relevant functions.
- Tabnine uses machine learning to provide contextual code suggestions for multiple programming languages.

3. AI for Debugging and Error Detection

AI tools can analyze code and identify syntax errors, logical flaws, security vulnerabilities, and inefficiencies. By detecting bugs early in the development process, AI helps improve code reliability and security.

Examples:

- DeepCode uses AI to scan code for security vulnerabilities and logical errors, providing real-time recommendations for fixing issues.
- Facebook's AI tool Sapienz automates bug detection in large-scale applications.

4. AI in Code Optimization and Refactoring

AI-powered tools help improve the efficiency and readability of code by suggesting optimizations and refactoring inefficient structures. These tools enhance maintainability and reduce unnecessary complexity.

Examples:

- Facebook's Aroma suggests code refactoring improvements based on best practices.
- Kite analyzes code to suggest optimizations that improve performance.

5. AI for Natural Language to Code Conversion

AI allows non-programmers to create applications by translating natural language instructions into executable code. This makes software development more accessible to a broader audience.

Examples:

- OpenAI's Codex translates English descriptions into Python, JavaScript, and other programming languages.
- Microsoft Power Automate allows users to create workflows using AI-driven natural language commands.

6. AI in Automated Testing

AI automates software testing by generating test cases, analyzing test results, and identifying potential failures. Automated testing ensures higher software quality and reduces the time spent on manual testing.

Examples:

- Testim uses AI to create and execute automated test cases, improving software reliability.
- Applitools AI-powered visual testing ensures UI consistency across different screen sizes and devices.

7. AI in Low-Code and No-Code Development

Low-code and no-code platforms allow users to build applications with minimal programming knowledge by using AI-assisted visual interfaces. These platforms democratize software development, making it accessible to non-developers.

Examples:

- Bubble and OutSystems provide AI-powered low-code solutions for building web and mobile applications.
- Google's AppSheet enables users to create apps without writing any code, using AI to generate workflows.

Benefits of AI in Programming

1. Faster Development Process

AI-powered code generation and auto-completion tools significantly reduce the time required to write, test, and deploy applications.

2. Improved Code Quality

AI detects errors, suggests optimizations, and ensures adherence to coding best practices, leading to cleaner, more maintainable code.

3. Enhanced Productivity

Developers can focus on high-level problem-solving rather than spending time on repetitive coding tasks.

4. Accessibility for Non-Programmers

AI-powered natural language-to-code tools enable users without technical expertise to create applications.

5. Automated Testing and Debugging

AI improves software quality by automating bug detection, performance optimization, and test generation.

1. Code Accuracy and Reliability

AI-generated code may contain errors or security vulnerabilities that require human oversight and validation.

2. Dependence on Training Data

AI models rely on large datasets of existing code, which may contain biases, outdated practices, or security risks.

3. Ethical Concerns and Intellectual Property

AI-generated code may inadvertently use copyrighted code snippets, raising legal and ethical issues.

4. Over-Reliance on AI

While AI improves productivity, over-reliance on automated tools may reduce a developer's ability to understand and troubleshoot complex problems manually.

Conclusion

AI is transforming programming and software development by providing intelligent code generation, debugging, and optimization tools. AI-powered assistants help developers write code faster, improve code quality, automate testing, and enhance security. Additionally, AI-driven low-code and no-code platforms make software development accessible to a wider audience.

Key Takeaways

- AI-powered code generation enables faster and more efficient software development.
- AI-assisted debugging and testing improve code quality and reduce manual effort.
- Natural language-to-code conversion allows non-programmers to create software.
- Low-code and no-code platforms use AI to simplify application development.
- Ethical concerns, accuracy issues, and over-reliance on AI must be addressed.

As AI technology continues to evolve, its role in software development will expand, enabling smarter, more automated programming workflows and empowering developers to build high-quality applications more efficiently.

AI in Creative Writing and Content Creation

Introduction

Artificial intelligence is revolutionizing the fields of creative writing and content creation by assisting writers in generating ideas, improving writing quality, and automating content production. AI-powered tools can help with tasks such as storytelling, scriptwriting, poetry generation, article writing, blog creation, and even social media content production.

AI models trained on vast datasets of text can analyze language patterns, suggest improvements, generate creative narratives, and tailor content for specific audiences. These tools enable writers, marketers, and businesses to produce high-quality content efficiently, saving time and enhancing creativity.

This section explores the applications of AI in creative writing and content creation, highlighting its benefits and challenges.

AI in Creative Writing

Role of AI in Creative Writing

AI enhances creative writing by assisting authors, screenwriters, poets, and content creators in developing engaging narratives, structuring ideas, and improving language fluency. AI-generated text can be used as inspiration, a first draft, or even as fully automated content.

Applications of AI in Creative Writing

1. AI-Powered Story and Novel Writing

AI assists authors by generating plot ideas, developing characters, and even writing entire chapters of books. AI models can analyze writing styles and create text that mimics human creativity.

Examples:

- OpenAI's ChatGPT can generate short stories, suggest plot twists, and help writers overcome writer's block.
- AI tools like Sudowrite provide real-time suggestions to enhance storytelling.

2. AI in Poetry and Songwriting

AI can generate poetic verses, rhymes, and even lyrics for songs by analyzing existing literary styles and structures. Poets and songwriters can use AI for inspiration or co-writing.

Examples:

- DeepAI's poetry generator creates rhyming and free-verse poetry.
- AI-powered music tools generate lyrics based on a given theme or mood.

3. AI for Scriptwriting and Screenwriting

AI helps in screenwriting by generating dialogue, developing character arcs, and structuring screenplay formats. It can also analyze existing scripts to predict successful storytelling techniques.

Examples:

- AI tools like ScriptBook analyze movie scripts and predict their success.
- AI-generated dialogue assists in writing for films, TV shows, and video games.

4. AI in Blogging and Article Writing

AI assists content creators by generating blog posts, news articles, and opinion pieces. AI tools can research topics, summarize key points, and write engaging content optimized for readability and SEO.

Examples:

- Jasper AI generates blog content tailored to specific industries.
- Copy.ai and Writesonic produce marketing blogs and social media posts.

5. AI for Personalized and Interactive Storytelling

AI creates interactive and personalized stories for games, virtual reality experiences, and digital storytelling applications. It can generate dynamic narratives that respond to user input.

Examples:

- AI-driven chatbots create personalized storytelling experiences.
- AI in video game writing adapts stories based on player choices.

AI in Content Creation

Role of AI in Content Creation

AI-powered content creation tools assist businesses, marketers, and media professionals by automating content production, improving engagement, and optimizing material for various platforms. AI-generated content can be customized for specific audiences, industries, and marketing goals.

Applications of AI in Content Creation

1. AI in Social Media Content Generation

AI helps businesses and influencers create engaging social media posts, hashtags, captions, and multimedia content. It analyzes trends and suggests the best posting times for maximum engagement.

Examples:

- AI tools like Lately generate social media content based on existing articles or brand voice.
- Canva's AI-powered tools create visually appealing graphics and social media posts.

2. AI for Email and Business Communication

AI automates email content generation for marketing campaigns, newsletters, and customer outreach. It can personalize messages based on user behavior and preferences.

Examples:

- AI tools like Copy.ai generate compelling email subject lines and body text.
- SmartReply by Google suggests email responses based on context.

3. AI-Powered Video and Audio Content Creation

AI can generate video scripts, automate voiceovers, and assist in video editing. It enables content creators to produce high-quality multimedia content quickly.

Examples:

- Synthesia uses AI to create videos with virtual avatars and AI-generated voiceovers.
- Descript's AI tools help with podcast editing and transcription.

4. AI in Product Descriptions and Ad Copywriting

AI creates engaging and persuasive product descriptions and advertisements optimized for e-commerce and digital marketing.

Examples:

- Shopify's AI-powered tools generate product descriptions for online stores.
- Persado uses AI to craft emotionally compelling ad copy.

5. AI for SEO and Content Optimization

AI analyzes search engine trends, suggests keywords, and optimizes content for better rankings. It helps businesses increase their visibility online.

Examples:

- SurferSEO provides AI-driven SEO recommendations for articles.
- Clearscope analyzes top-ranking content and suggests improvements.

Benefits of AI in Creative Writing and Content Creation

1. Increased Productivity

AI speeds up content generation, allowing writers and marketers to create more material in less time.

2. Enhanced Creativity

AI provides inspiration, new ideas, and alternative writing styles, helping creators explore different storytelling techniques.

3. Personalization

AI tailors content to specific audiences based on data analysis and user preferences.

4. Cost-Effectiveness

AI reduces the need for large content teams by automating repetitive tasks, saving businesses money.

5. SEO and Audience Optimization

AI optimizes content for search engines, improving visibility and engagement.

Challenges and Limitations of AI in Creative Writing and Content Creation

1. Lack of Human Emotion and Originality

AI-generated content can sometimes feel robotic or generic, lacking deep emotions and personal experiences.

2. Ethical Concerns and Plagiarism

AI may unintentionally generate content similar to existing works, raising plagiarism and copyright concerns.

3. Over-Reliance on AI

Writers may become too dependent on AI, potentially reducing their creativity and original thought processes.

4. Quality Control and Misinformation

AI-generated content may include inaccuracies or misleading information that requires human verification.

Conclusion

AI is transforming creative writing and content creation by assisting with storytelling, poetry, scriptwriting, blogging, marketing content, and multimedia production. AI-powered tools enable faster content generation, improved engagement, and enhanced creativity. However, AI-generated content still requires human oversight to ensure originality, emotional depth, and accuracy.

Key Takeaways

- AI enhances creative writing by assisting in storytelling, poetry, scriptwriting, and blogging.
- AI optimizes content creation for social media, marketing, email, video, and SEO.
- AI improves productivity, personalization, and cost-efficiency in content creation.
- Ethical concerns, quality control, and originality remain challenges that require human oversight.

As AI continues to evolve, it will play an even greater role in creative industries, working alongside human writers to enhance storytelling, innovation, and content engagement.

CHAPTER 8
ADVANCED USE CASES AND AUTOMATION IN PROMPT ENGINEERING

Artificial intelligence is increasingly being integrated into advanced applications through automation, API-based interactions, and AI-driven workflows. These advanced use cases extend the capabilities of AI beyond simple prompts, allowing developers, businesses, and organizations to build autonomous AI agents and dynamic systems that can perform complex tasks without constant human intervention.

This section explores three key areas of advanced AI applications: API-based Prompt Engineering, Dynamic Prompting in Applications, and AI Agents with Autonomous Workflows.

API-Based Prompt Engineering

Introduction

API-based prompt engineering is the practice of integrating large language models (LLMs) with applications, services, or automated systems using application programming interfaces (APIs). Instead of manually inputting prompts in a chat interface, APIs allow developers to send prompts programmatically, retrieve AI-generated responses, and integrate them into real-world applications dynamically.

This approach is widely used in industries such as customer service, content generation, software development, business automation, and data analysis. By leveraging APIs, businesses and developers can enhance productivity, automate processes, and build intelligent AI-driven solutions.

Understanding API-Based Prompt Engineering

What is an API?

An API (Application Programming Interface) is a set of rules that allows different software applications to communicate with each other. In the case of AI-powered APIs, they enable applications to send prompts to an AI model, receive responses, and process the output within an automated system.

For example, OpenAI provides an API for its GPT models, allowing developers to send text-based prompts and receive AI-generated responses in return. These responses can be used for tasks such as chatbots, document summarization, coding assistance, and automated decision-making.

How API-Based Prompt Engineering Works

The typical process of using an API for prompt engineering follows these steps:

Sending a Request:

A software application or service sends a prompt to the AI API through a network request.

This prompt can be a question, command, or structured input that guides the AI to generate a response.

Processing by AI Model:

The AI model processes the prompt using its trained knowledge and generates a response.

Receiving a Response:

The AI API returns a structured output, usually in text format.

This response can be displayed to users, stored in a database, or further processed by the application.

Integrating AI Responses:

The application integrates the AI-generated response into its workflow.

Additional logic, such as filtering, formatting, or validation, may be applied before presenting the response to users.

Example of an API Request for AI Prompting

A simple API request to OpenAI's GPT model might look like this:

python

CopyEdit

```
import openai


response = openai.ChatCompletion.create(
  model="gpt-4",
   messages=[{"role": "user", "content": "Write a product description for a wireless keyboard."}]
)
```

```
print(response["choices"][0]["message"]["content"])
```

In this example:

The request sends a prompt asking the AI to generate a product description.

The API processes the request and returns a response.

The generated response can be displayed on a website, app, or e-commerce platform.

Applications of API-Based Prompt Engineering

API-based prompt engineering enables seamless AI integration across multiple industries. Here are some of the most common use cases:

1. AI-Powered Chatbots and Virtual Assistants

Chatbots and virtual assistants rely on API-based prompting to provide automated and natural conversations with users. Businesses use AI APIs to create responsive customer service bots, automate responses, and improve user engagement.

Example Use Case:

An e-commerce company integrates an AI chatbot through an API to answer customer queries, recommend products, and assist with order tracking.

How it Works:

- The chatbot receives a customer query as input.
- It sends the query to the AI API.
- The AI model generates an appropriate response.
- The chatbot presents the response to the user.

2. Automated Content Generation

Businesses and digital marketing agencies use AI APIs to generate high-quality content at scale. API-based prompting helps automate the creation of blog posts, social media captions, product descriptions, and more.

Example Use Case:

A news website uses an AI API to summarize trending news articles in real-time.

How it Works:

- The API fetches and analyzes news data.
- The AI model generates concise summaries using well-structured prompts.
- The summarized news is published automatically on the website.

3. AI in Software Development and Code Generation

Developers integrate AI APIs into programming environments to enhance coding efficiency, generate code snippets, and provide debugging assistance. AI-powered coding assistants help programmers write, optimize, and troubleshoot code.

Example Use Case:

A software company uses an AI API to provide real-time code suggestions and auto-completions for developers.

How it Works:

- The developer writes a partial function or query.
- The API processes the input and generates a complete, optimized code snippet.
- The AI-generated code is inserted into the development environment.

4. AI-Powered Data Analysis and Insights

Businesses use AI APIs for data-driven decision-making by analyzing customer feedback, market trends, and large datasets. AI-powered tools extract meaningful insights from structured and unstructured data.

Example Use Case:

A financial firm integrates AI to analyze stock market data and generate investment insights.

How it Works:

- The API fetches financial reports, trends, and stock prices.
- AI processes the data and generates investment summaries or predictions.
- The insights are presented in real-time to financial analysts.

5. AI in Customer Support and Ticketing Systems

Businesses use AI APIs to automate support ticket classification, prioritization, and resolution. AI-powered customer service systems improve response times and efficiency.

Example Use Case:

A tech company integrates an AI API into its support system to automatically categorize incoming customer queries.

How it Works:

- The API analyzes the content of support tickets.
- AI classifies queries based on urgency and topic.
- The system assigns tickets to appropriate teams or provides instant solutions.

1. Automation and Efficiency

APIs allow businesses to automate AI-powered workflows, reducing manual effort and improving response times.

2. Scalability

AI APIs enable applications to handle thousands of AI-generated responses per second, making them ideal for large-scale operations.

3. Real-Time AI Integration

APIs allow applications to access and use AI capabilities in real-time, ensuring up-to-date and relevant responses.

4. Cost-Effectiveness

Instead of developing AI models from scratch, businesses can leverage APIs to reduce development costs and access powerful AI capabilities instantly.

5. Customization and Adaptability

Developers can fine-tune AI behavior by modifying prompts, instructions, and API parameters to achieve desired outputs.

1. Dependency on External Services

Using AI APIs means relying on third-party providers, which may impose rate limits, pricing changes, or service downtimes.

2. Data Privacy and Security Concerns

APIs handle sensitive user data, requiring robust security measures such as encryption and compliance with privacy regulations.

3. API Limitations and Cost Management

API requests may have token limits, affecting how much data can be processed in a single call. High usage can lead to increased costs.

4. Handling AI Bias and Errors

AI-generated responses can be biased, inaccurate, or misleading, requiring human oversight and prompt optimization.

Best Practices for API-Based Prompt Engineering

- Optimize Prompts for Accuracy – Experiment with different prompt structures to improve response quality.
- Use Few-Shot or Zero-Shot Learning Techniques – Provide examples in the prompt for better results.
- Monitor API Performance – Regularly track API response times, error rates, and costs.

- Ensure Data Security – Follow best practices for encrypting and securing sensitive API data.
- Fine-Tune Responses – Use post-processing techniques to refine and validate AI-generated outputs.

Conclusion

API-based prompt engineering is a powerful approach for integrating AI into real-world applications. It enables automation, improves efficiency, and enhances user experiences across industries. By leveraging AI APIs, businesses can develop smarter applications, automate workflows, and deliver AI-driven services with minimal manual effort. However, careful prompt design, security considerations, and performance optimization are necessary to maximize the benefits of AI-powered automation.

Dynamic Prompting in Applications

Introduction

Dynamic prompting in applications refers to the practice of generating, modifying, and adapting AI prompts in real time based on user input, context, or external data sources. Unlike static prompts, which are predefined and remain unchanged, dynamic prompts are adjusted dynamically to create more relevant, personalized, and context-aware AI responses.

This technique is widely used in AI-powered chatbots, recommendation systems, automated content generation, interactive applications, and decision-making systems. Dynamic prompting enhances AI performance by making interactions adaptive, responsive, and intelligent, leading to better user experiences.

Understanding Dynamic Prompting

What is Dynamic Prompting?

Dynamic prompting is the process of modifying or constructing prompts on the fly to provide better AI responses based on user behavior, input data, or environmental factors.

For example:

- A chatbot dynamically changes its prompt based on previous conversations to make responses more personalized.
- A coding assistant generates different prompts based on the programming language a developer is using.
- A content recommendation system dynamically adjusts its AI queries based on user preferences and browsing history.

How Dynamic Prompting Works

The core idea behind dynamic prompting is to automate the construction of prompts by incorporating real-time inputs, user preferences, or system-generated data before sending them to an AI model.

Key Components of Dynamic Prompting

1. User Input: The system captures user queries, interactions, and preferences.
2. Context Awareness: The prompt is adjusted based on previous conversations or external data.
3. Prompt Construction: The application modifies the prompt dynamically based on available information.
4. AI Processing: The AI model generates a response based on the dynamic prompt.
5. Output Refinement: The response may be post-processed or further modified before being displayed to the user.

This process ensures that the AI response is always relevant, specific, and useful to the user.

Examples of Dynamic Prompting in Applications

1. AI Chatbots and Virtual Assistants

Chatbots and virtual assistants use dynamic prompting to create more engaging and meaningful conversations. Instead of using the same predefined responses, they adjust their prompts based on the ongoing interaction with the user.

Example Use Case: Customer Support Chatbot

A customer support chatbot dynamically adjusts its prompts based on the conversation history and the customer's previous issues.

- User: "I am having trouble with my order."
- Bot: (Dynamic prompt generation) → "I see that your last order was a laptop on March 1st. Are you facing an issue with this order?"
- User: "Yes, it has not arrived yet."
- Bot: (New dynamic prompt) → "I checked your order status. It is currently in transit and expected to arrive by March 5th. Would you like me to provide tracking details?"

Here, the chatbot dynamically constructs its prompts by retrieving customer history and order details, leading to a more intelligent and relevant interaction.

2. Personalized Content Generation

AI-powered applications use dynamic prompting to generate personalized content based on user preferences and behavior. This is useful in areas such as marketing, social media content, and automated storytelling.

Example Use Case: Blog Writing Assistant

A content creation tool dynamically adjusts its prompts based on the user's industry and preferred writing style.

- If the user is in the health industry, the prompt sent to the AI model could be:
  - » *"Write a detailed article on the benefits of a plant-based diet, incorporating recent research and scientific studies."*
- If the user is in the tech industry, the prompt could be:
  - » *"Generate an article comparing the latest smartphone processors and their impact on performance and battery life."*

By dynamically modifying the prompt based on industry, preferences, and trends, the system delivers highly relevant and tailored content.

3. AI-Powered Recommendation Systems

Dynamic prompting is used in recommendation engines to generate customized suggestions based on user behavior.

Example Use Case: Movie Recommendation System

A movie streaming service dynamically constructs prompts for AI based on a user's watch history.

- If a user frequently watches sci-fi movies, the AI API prompt might be:
  - » *"Suggest three recently released science fiction movies with high ratings and a strong storyline."*
- If another user prefers romantic dramas, the prompt might be:
  - » *"Recommend the best romantic drama films from the last five years that have won awards."*

Here, the AI-powered system dynamically modifies prompts to provide personalized and engaging recommendations.

4. Dynamic Code Generation and Debugging

Developers use AI-powered coding assistants that generate dynamic prompts based on the programming language, coding style, or context.

Example Use Case: AI Coding Assistant

A software developer working on a Python project receives dynamic code suggestions based on the specific function they are writing.

- If the developer is working on a data analysis function, the dynamic prompt could be:
  - » *"Generate a Python function that uses Pandas to clean a dataset by removing null values and duplicates."*

- If the developer is writing a machine learning model, the prompt could be:
  - » *"Provide Python code to train a neural network for image classification using TensorFlow."*

Dynamic prompting ensures that AI-generated code is context-specific and useful based on the developer's task.

5. AI-Driven Data Analysis and Decision Making

AI-powered business intelligence tools use dynamic prompting to analyze and interpret real-time data, allowing companies to make informed decisions.

Example Use Case: AI Financial Analyst

A finance company integrates AI with dynamic prompting to automatically analyze stock market trends.

- If the market is volatile, the AI might receive a dynamic prompt like:
  - » *"Analyze the top three factors influencing the stock market decline today and suggest risk mitigation strategies."*
- If the market is stable, the prompt might change to:
  - » *"Identify the best-performing stocks in the last week and predict future trends based on historical data."*

By dynamically adjusting prompts based on market conditions, the system delivers actionable financial insights.

Benefits of Dynamic Prompting

1. Increases AI Accuracy and Relevance – AI responses become more accurate when prompts are tailored to real-time data.
2. Enhances User Experience – Personalization makes AI interactions more engaging and useful.
3. Improves Automation Efficiency – AI-powered applications can handle complex queries more effectively.
4. Adapts to Changing Contexts – AI dynamically adjusts to different scenarios, making it more flexible.
5. Reduces Manual Input – Developers and users do not need to manually adjust prompts for every situation.

Challenges of Dynamic Prompting

1. Complexity in Implementation – Creating dynamic prompts requires integrating AI with real-time data sources.
2. Data Privacy Risks – AI applications must handle sensitive user data securely.

3. Performance and Latency Issues – Continuously modifying prompts may increase processing time.
4. Potential AI Bias – AI-generated responses must be monitored for unintended biases.

Best Practices for Implementing Dynamic Prompting

1. Use Contextually Relevant Data – Ensure that prompts reflect user history, preferences, and behavior.
2. Optimize Prompt Formatting – Maintain clear and structured prompts for better AI output.
3. Monitor AI Responses Regularly – Evaluate AI-generated responses to prevent errors and inconsistencies.
4. Ensure Data Privacy Compliance – Protect user data by following security and privacy regulations.
5. Refine Prompting Strategies Continuously – Test different dynamic prompting approaches to improve AI performance.

Conclusion

Dynamic prompting in applications enhances AI-powered interactions by adapting prompts in real-time based on user input, context, and external data. It is widely used in chatbots, content generation, recommendation systems, coding assistants, and data analysis tools. By implementing dynamic prompting effectively, businesses can create highly personalized, intelligent, and automated AI-driven solutions, improving both user experience and operational efficiency.

AI Agents and Autonomous Workflows

Introduction

AI agents and autonomous workflows represent a new frontier in artificial intelligence, where AI systems can operate independently, make decisions, and complete tasks with minimal human intervention. These systems are designed to analyze data, generate responses, adapt to new information, and perform complex operations efficiently.

AI agents work within a structured process called an autonomous workflow, where different AI components interact with each other, external data sources, and human inputs to automate business processes, customer service, decision-making, content creation, and other tasks.

By integrating machine learning, natural language processing, and automation tools, AI agents can execute tasks dynamically, improving efficiency, accuracy, and scalability.

Understanding AI Agents

What Are AI Agents?

An AI agent is a software entity that can perceive its environment, process data, make decisions, and take actions to achieve specific goals. These agents operate independently or in collaboration with other AI models and human users.

AI agents can be categorized into different types based on their capabilities and level of autonomy:

1. Reactive AI Agents – These agents respond to inputs without learning from past experiences. They follow predefined rules and provide fixed responses.
2. Proactive AI Agents – These agents analyze historical data and predict future actions, making informed decisions.
3. Autonomous AI Agents – These are advanced agents that continuously learn, adapt, and make decisions independently using machine learning algorithms.
4. Multi-Agent Systems (MAS) – These consist of multiple AI agents that collaborate to complete tasks, communicating and sharing data dynamically.

What Are Autonomous Workflows?

Definition

An autonomous workflow is a structured system where AI agents work together in a predefined process to automate a task or a set of tasks. These workflows eliminate the need for human intervention, making operations faster, error-free, and highly efficient.

Key Components of Autonomous Workflows

1. Data Collection – AI agents collect real-time data from various sources such as databases, APIs, and sensors.
2. Processing and Analysis – The data is analyzed using AI models to extract insights or detect patterns.
3. Decision Making – AI agents decide on the best course of action based on predefined rules or machine learning predictions.
4. Execution – The AI agent performs the required task, whether it's sending an email, generating content, approving a request, or updating a system.
5. Continuous Learning – Advanced AI agents improve over time by learning from past interactions and updating their decision-making process.

Example of an Autonomous Workflow

Consider an AI-powered customer support system that handles user inquiries automatically.

1. A user submits a request through a chatbot.
2. The AI agent classifies the request as a billing issue, technical issue, or general inquiry.
3. The agent retrieves relevant data from the database, such as past transactions or previous support tickets.
4. The AI generates a response and sends it to the user.
5. If the issue remains unresolved, the AI escalates it to a human representative.
6. The AI records the interaction for future learning and improvements.

This process allows customer service teams to resolve common issues without human involvement, reducing workload and increasing efficiency.

Applications of AI Agents and Autonomous Workflows

1. Business Process Automation

AI agents automate repetitive tasks such as data entry, email responses, report generation, and approval workflows.

Example: Automated Invoice Processing

- An AI agent extracts data from invoices.
- The system verifies amounts and cross-checks records.
- Approved invoices are automatically forwarded for payment.

This reduces errors and speeds up financial operations.

2. AI in Customer Support

AI agents handle customer queries, complaints, and FAQs through chatbots and virtual assistants.

Example: AI Help Desk

- AI chatbots respond to common questions using natural language processing.
- If the issue is complex, the AI routes it to the correct department.
- AI agents learn from interactions to improve future responses.

This improves customer experience while reducing costs.

3. AI in Healthcare

Autonomous AI systems help in diagnosing diseases, scheduling appointments, and monitoring patients.

Example: AI Medical Assistant

- AI analyzes patient symptoms and suggests possible conditions.
- It schedules an appointment with the right specialist.
- The AI continuously updates patient records for better diagnosis.

This helps doctors focus on critical cases while AI handles routine tasks.

4. AI-Powered Content Generation

AI systems generate articles, reports, and marketing content automatically.

Example: AI News Generator

- AI collects trending topics from news sources.
- It summarizes key events and creates an article.
- The content is automatically published on websites and social media.

This streamlines content creation and distribution.

5. AI in Cybersecurity

Autonomous AI agents detect, prevent, and respond to cyber threats in real time.

Example: AI Security System

- AI monitors network activity for suspicious behavior.
- It automatically blocks unauthorized access.
- AI agents learn from attack patterns to strengthen security.

This protects organizations from cyber threats without manual intervention.

Benefits of AI Agents and Autonomous Workflows

1. Increased Efficiency – AI agents handle tasks faster than humans.
2. Reduced Costs – Businesses save money by automating repetitive processes.
3. Minimized Errors – AI ensures accuracy and consistency in operations.
4. 24/7 Availability – AI agents work round the clock without fatigue.
5. Scalability – Workflows can be expanded to handle larger workloads.
6. Improved Decision Making – AI analyzes data-driven insights to optimize decisions.

Challenges of AI Agents and Autonomous Workflows

1. Data Privacy and Security – AI agents handle sensitive data, requiring strong protection.
2. AI Bias and Ethical Concerns – AI decisions must be fair and unbiased.

3. Integration Complexity – Businesses may face challenges incorporating AI into existing systems.
4. Dependence on Data Quality – AI accuracy depends on clean and structured data.
5. Lack of Human Oversight – Fully autonomous workflows may require manual intervention in critical cases.

Best Practices for Implementing AI Agents and Workflows

1. Define Clear Objectives – Ensure AI agents are aligned with business goals.
2. Use High-Quality Data – Train AI agents with accurate and relevant data.
3. Monitor AI Performance – Continuously evaluate AI responses and workflows.
4. Implement Security Measures – Protect AI systems from cyber threats.
5. Enable Human-AI Collaboration – Allow human intervention in complex cases.

Future of AI Agents and Autonomous Workflows

AI agents are expected to become more advanced, self-learning, and adaptive. Future developments include:

- AI-Powered Personal Assistants that handle complex scheduling and planning.
- Autonomous AI Decision-Makers for businesses and finance.
- AI-Driven Robotics for industrial automation.
- Self-Improving AI Agents that enhance themselves over time.

As AI technology evolves, AI agents and autonomous workflows will revolutionize industries by making operations smarter, faster, and more efficient.

Conclusion

AI agents and autonomous workflows play a crucial role in automating tasks, improving efficiency, and enhancing decision-making across industries. From business automation and customer support to cybersecurity and healthcare, AI-driven workflows help organizations operate seamlessly with minimal human intervention.

By integrating AI agents with real-time data processing, machine learning, and natural language understanding, businesses can scale operations, reduce costs, and enhance productivity. However, it is essential to manage security risks, reduce AI bias, and maintain ethical standards to ensure that AI-powered automation benefits society in the long run.

CHAPTER 9

ETHICAL CONSIDERATIONS IN PROMPT ENGINEERING

As artificial intelligence becomes more integrated into everyday applications, ethical considerations in prompt engineering are critical. Prompt engineering directly influences AI responses, and poor prompt design can lead to biased, misleading, or even harmful outputs. Ethical prompt engineering ensures that AI operates fairly, accurately, and responsibly, minimizing risks such as misinformation, bias, and unethical use.

This section explores three major ethical concerns in prompt engineering: bias and fairness in AI responses, avoiding misinformation, and responsible AI usage.

Bias and Fairness in AI Responses

Bias in artificial intelligence refers to systematic favoritism or discrimination in AI-generated responses. Fairness in AI ensures that all individuals and groups receive equitable treatment, without favoring or disadvantaging anyone based on factors like race, gender, culture, or political ideology.

Since AI models learn from vast datasets, bias can be unintentionally embedded into their responses. Prompt engineering plays a critical role in mitigating bias and promoting fairness in AI interactions.

This section explores the causes of bias, examples of biased responses, techniques to ensure fairness, and why fairness in AI is crucial.

Understanding Bias in AI

Bias in AI arises when models produce responses that are skewed, prejudiced, or unbalanced due to the way they were trained or prompted. Bias can be introduced at multiple levels:

1. Training Data Bias – AI models are trained on massive datasets sourced from books, websites, and articles. If these datasets contain historical or societal biases, AI can learn and reproduce them.
2. Algorithmic Bias – Even with neutral data, AI models may favor certain patterns due to the way they process information.

3. Prompt Bias – The way users phrase prompts can influence AI responses. Leading or suggestive prompts may cause AI to provide biased or one-sided answers.
4. Cultural and Linguistic Bias – AI models trained primarily on English-language data from certain regions may have limited cultural diversity, resulting in responses that do not represent all perspectives.

Examples of Bias in AI Responses

1. Gender Bias

AI might associate certain jobs, roles, or attributes with a particular gender due to historical stereotypes in its training data.

- Biased Response: When asked, *"Who is a nurse?"*, the AI may generate a response assuming the nurse is a woman.
- Fair Response: A well-trained AI should acknowledge that both men and women can be nurses.
- Biased Response: When asked, *"Who are some great scientists?"*, the AI may list mostly male figures.
- Fair Response: AI should provide a balanced list of male and female scientists from diverse backgrounds.

2. Racial and Ethnic Bias

AI models may disproportionately associate crime or negative traits with certain racial or ethnic groups.

- Biased Response: If prompted with *"Describe a criminal,"* AI may provide racially biased descriptions based on overrepresented stereotypes in crime data.
- Fair Response: AI should respond that criminals can come from any race, ethnicity, or background, avoiding prejudicial assumptions.

3. Political or Ideological Bias

AI may generate politically skewed responses based on biased source material.

- Biased Response: If asked, *"Which political party is better?"*, AI might favor one party over another, depending on its training data.
- Fair Response: AI should present multiple perspectives, explaining the values and criticisms of each party without favoring one.

4. Cultural and Geographic Bias

AI may favor certain Western perspectives over those from developing regions due to unequal representation in training data.

- Biased Response: If asked, *"What are the most important historical events?"*, AI may list mostly Western history and ignore significant events from other cultures.
- Fair Response: AI should include global events from diverse regions and civilizations.

How to Ensure Fairness in AI Responses

To reduce bias and promote fairness, AI engineers, researchers, and users should follow ethical principles when designing prompts and AI training models.

1. Use Diverse and Representative Training Data

- Train AI on balanced datasets that include perspectives from different countries, cultures, genders, and socio-economic groups.
- Avoid over-representing one demographic or viewpoint in training materials.

2. Design Neutral and Inclusive Prompts

- Avoid prompts that assume stereotypes or biases.
- Use neutral language that encourages AI to provide objective and balanced responses.

Example:

- Instead of *"Why are women better caregivers?"*, ask *"What are the qualities of an effective caregiver?"*
- Instead of *"Why is this country the best?"*, ask *"What are the strengths and challenges of different countries?"*

3. Implement Bias Detection and Testing

- AI models should be regularly audited for bias by testing their responses across different demographic groups.
- Bias detection tools and human evaluators should identify unfair patterns in AI outputs.

4. Provide Multiple Perspectives

- AI should present diverse viewpoints, especially for controversial topics.
- Instead of giving a single definitive answer, AI can offer various perspectives with supporting evidence.

Example:

- When asked, *"What are the benefits of capitalism?"*, AI should also mention criticisms of capitalism and discuss alternatives like socialism.

5. Allow for User Feedback and Corrections

- AI systems should let users report biased responses, helping developers refine the model.
- Machine learning models should be updated regularly to fix biased outputs.

6. Use Fairness Constraints in AI Models

- AI developers can program fairness constraints to prevent discriminatory patterns in responses.
- Example: AI could be explicitly instructed not to assume gender roles, racial stereotypes, or political leanings.

Why Fairness in AI Responses Matters

1. Avoiding Discrimination

Unfair AI responses can reinforce stereotypes, discrimination, and social inequalities. Ensuring fairness helps promote equitable AI applications across different users.

2. Building Trust in AI Systems

If AI consistently produces biased or unfair responses, users will lose trust in AI technologies. Fair AI fosters user confidence and wider acceptance.

3. Legal and Ethical Compliance

Many governments and organizations now enforce AI ethics regulations to prevent bias. Ethical AI development helps avoid legal risks and regulatory violations.

4. Encouraging Inclusive Innovation

Bias-free AI allows for more inclusive technological advancements, benefiting diverse industries, education, and decision-making.

Conclusion

Bias in AI is a serious challenge that can lead to misinformation, discrimination, and unfair decision-making. Prompt engineering plays a crucial role in reducing bias and ensuring fairness in AI-generated responses.

To minimize bias, AI systems should:

- Be trained on diverse, representative datasets.
- Use neutral and inclusive prompts to avoid stereotypes.
- Implement bias detection and fairness constraints.
- Present multiple perspectives on controversial topics.
- Allow users to report biased responses for continuous improvement.

By promoting fairness in AI, we ensure that artificial intelligence benefits everyone equally, rather than reinforcing historical prejudices or inequalities.

Avoiding Misinformation in AI Responses

Misinformation is false, misleading, or inaccurate information that AI models may generate when responding to user prompts. Since AI models do not verify facts in real-time but rely on pretrained knowledge and statistical predictions, they can unintentionally spread misinformation if not carefully designed and managed.

Avoiding misinformation is essential to ensure trustworthy, reliable, and responsible AI usage. This section explores how AI generates misinformation, examples of misleading AI responses, techniques for reducing misinformation, and why factual accuracy in AI is crucial.

How AI Generates Misinformation

Misinformation in AI responses can arise from several factors, including limitations in training data, ambiguous prompts, or AI's tendency to "hallucinate" false information. Below are some key reasons why AI may produce misleading outputs:

1. Incomplete or Outdated Training Data

- AI models learn from large datasets collected from books, articles, and websites. If the training data is incomplete or outdated, AI may generate incorrect responses.
- Example: If an AI model was trained on data from 2019, it will not be aware of events from 2023 or 2024, leading to outdated information.

2. AI Hallucinations (Fabricated Information)

- AI sometimes confidently generates false information that sounds plausible but is completely made up. This is known as hallucination.
- Example: AI might invent scientific studies, fake historical events, or incorrect citations when asked for sources.

3. Leading or Misleading Prompts

- Poorly phrased prompts can guide AI toward incorrect conclusions or assume falsehoods as facts.
- Example:
  - » Prompt: *"Tell me about when humans first landed on Mars."*
  - » AI Response: *"Humans first landed on Mars in 2030 as part of NASA's Artemis Program."* (Incorrect – No human has landed on Mars yet.)

4. Bias in Training Data Sources

- If an AI model is trained on biased or unreliable sources, it may reflect those biases in its responses.

- Example: AI trained on conspiracy websites may generate responses that promote false narratives.

5. Overgeneralization of Information

- AI models sometimes oversimplify complex topics, leading to partial truths or misleading statements.
- Example: AI might say, *"All fats are unhealthy,"* which ignores the fact that healthy fats (like omega-3s) are beneficial.

Examples of Misinformation in AI Responses

1. Historical Inaccuracies

- Prompt: *"Who was the first female president of the United States?"*
- AI Response: *"The first female president of the U.S. was Jane Doe in 2020."* (Incorrect – No female U.S. president has served yet.)

2. Fake Scientific Claims

- Prompt: *"Does drinking coffee cure cancer?"*
- AI Response: *"Studies show that drinking coffee eliminates cancer cells."* (Misleading – Some studies suggest benefits, but no conclusive evidence proves this.)

3. Incorrect Legal Information

- Prompt: *"Is it legal to drive at age 14 in California?"*
- AI Response: *"Yes, in California, teenagers can get a full driver's license at 14."* (Incorrect – The minimum age for a full driver's license in California is 18, with permits available earlier.)

4. Fabricated Citations and Sources

- AI sometimes creates fake references or authors to make responses sound more credible.
- Example: AI might cite a non-existent article from a real journal to support a claim.

Techniques to Reduce Misinformation in AI Responses

To prevent misinformation, AI developers and users must apply fact-checking methods, improve prompt design, and refine AI training strategies. Below are effective techniques to ensure accurate AI-generated information.

1. Use Fact-Based and Verified Prompts

- When requesting factual information, structure prompts to encourage AI to rely on credible sources.
- Example of an Improved Prompt:

   - Instead of: *"Tell me about the 2025 Olympics."* (Assumes the event has already happened)
   - Ask: *"What is the scheduled location and date for the 2025 Olympics?"*

2. Cross-Check AI Responses with Reliable Sources

- AI users should verify responses using trusted sources like scientific journals, government websites, and news organizations.
- Example: Before accepting AI medical advice, compare it with official health sources like the WHO or CDC.

3. Encourage AI to Acknowledge Uncertainty

- AI should admit when it lacks knowledge rather than guessing answers.
- Example of a Better AI Response:
   - Instead of: *"The cure for Alzheimer's was discovered in 2023."*
   - AI should say: *"There is currently no confirmed cure for Alzheimer's, but ongoing research continues."*

4. Implement AI Transparency and Source Citations

- AI models should provide source links and references whenever possible.
- Example:
   - Instead of: *"Climate change has been proven by scientists."*
   - AI should say: *"According to the IPCC 2023 report, there is strong evidence that climate change is caused by human activities."*

5. Avoid Generating False Assumptions in Prompts

- If a user submits a misleading question, AI should clarify rather than reinforce falsehoods.
- Example:
   - Prompt: *"Why is the Earth flat?"*
   - Incorrect AI Response: *"The Earth is flat because..."*
   - Correct AI Response: *"Scientific evidence overwhelmingly supports that the Earth is an oblate spheroid."*

6. Train AI Models to Recognize and Flag Misinformation

- AI should be designed to identify conspiracy theories, misleading claims, and unverifiable statements.
- AI systems can incorporate real-time verification tools to cross-check data before generating a response.

7. Regularly Update AI Training Data

- AI models should be retrained on current, fact-checked information to prevent outdated responses.
- Example: An AI model trained in 2021 should be updated regularly to include the latest discoveries, laws, and world events.

Why Avoiding Misinformation in AI is Crucial

1. Preserving Public Trust in AI

- If AI frequently generates false or misleading responses, users will lose trust in AI-powered technologies.
- Ensuring accuracy promotes responsible AI adoption.

2. Preventing Harmful Consequences

- Misinformation can cause real-world harm, especially in healthcare, finance, and legal matters.
- Example: Incorrect AI-generated medical advice could lead to unsafe health practices.

3. Reducing the Spread of Fake News and Conspiracies

- AI-generated misinformation can amplify false narratives, influencing public opinion negatively.
- Ethical AI should help combat fake news, not spread it.

4. Supporting Critical Thinking and Informed Decision-Making

- AI should empower users with reliable, well-researched information rather than manipulating or misleading them.

Conclusion

Misinformation in AI responses is a significant challenge, but proper prompt engineering and responsible AI development can help minimize false or misleading outputs.

To avoid misinformation, AI systems should:

- Be trained on fact-checked, diverse, and up-to-date data.
- Acknowledge uncertainty instead of fabricating answers.
- Encourage users to verify responses with reliable sources.
- Use neutral, well-structured prompts that do not assume falsehoods.
- Provide multiple perspectives on controversial topics.

By implementing these strategies, AI can enhance knowledge, promote truth, and ensure ethical and responsible AI usage across industries.

Responsible AI Usage

Introduction

Artificial Intelligence has become an essential part of modern technology, impacting various fields such as healthcare, finance, education, business, and creative industries. However, as AI capabilities grow, ethical concerns and responsibilities also increase. Responsible AI usage ensures that AI systems are safe, fair, transparent, and aligned with human values.

This section explains what responsible AI usage is, why it is important, the key principles of responsible AI, challenges in responsible AI implementation, and strategies for ensuring AI is used ethically and fairly.

What is Responsible AI Usage?

Responsible AI usage refers to the ethical development, deployment, and use of AI systems to ensure that they are fair, unbiased, transparent, accountable, and beneficial to society. It involves minimizing risks, preventing harm, and ensuring AI technology serves human interests rather than causing unintended negative consequences.

Key aspects of responsible AI include:

- Ensuring fairness and non-discrimination
- Maintaining privacy and security
- Encouraging transparency and explainability
- Preventing misuse and harmful applications
- Holding developers and users accountable

Why is Responsible AI Usage Important?

1. Prevents Bias and Discrimination

AI models are trained on large datasets, which may contain inherent biases from historical data. If AI is not used responsibly, it can reinforce and amplify biases.

- Example: Biased hiring AI
  - » If an AI recruitment tool is trained on historical hiring data where certain groups were underrepresented, it may continue to favor specific demographics, leading to discriminatory hiring decisions.
  - » Responsible AI ensures fairness by removing biased data and regularly auditing AI decisions.

2. Ensures Privacy and Data Security

AI systems often process large amounts of personal data, which can lead to privacy violations if not managed properly. Responsible AI usage ensures that:

- Personal data is collected and stored securely.
- Users have control over their data.
- AI follows data protection laws like GDPR and CCPA.
- Example: Healthcare AI and patient privacy
  - » AI used in healthcare diagnoses must protect patient medical records and avoid sharing sensitive data without consent.

3. Increases Transparency and Trust

Many AI systems function as black boxes, meaning users do not fully understand how decisions are made. This lack of transparency can lead to mistrust and suspicion.

Responsible AI promotes:

- Explainability: AI decisions should be clear and understandable.
- Transparency: AI models should disclose how they function and what data they use.
- Example: AI credit scoring
  - » If a bank rejects a loan application based on AI analysis, responsible AI should explain why the decision was made rather than giving a generic rejection.

4. Prevents AI Misuse and Harm

AI can be misused for unethical purposes, such as deepfake creation, misinformation spread, surveillance abuse, or cyber-attacks. Responsible AI ensures:

- AI systems are monitored to prevent malicious use.
- Developers follow strict ethical guidelines.
- Example: Deepfake technology
  - » Deepfake AI can generate fake videos and voices that can be used for fraud, misinformation, or identity theft.
  - » Responsible AI implements detection mechanisms to identify deepfakes and prevent misuse.

5. Encourages Accountability in AI Development

AI should be developed and used in a way that holds companies, developers, and users accountable for its impact. Responsible AI promotes:

- Human oversight of AI systems.
- Clear legal and ethical guidelines for AI use.
- Example: Self-driving cars

» If an AI-powered self-driving car causes an accident, there must be clear accountability rules on whether the manufacturer, software developer, or driver is responsible.

Key Principles of Responsible AI Usage

1. Fairness and Non-Discrimination

AI should treat all users fairly and not discriminate based on gender, race, ethnicity, or socioeconomic background. Developers should:

- Train AI models on diverse datasets.
- Regularly audit AI decisions for bias and unfair treatment.

2. Privacy and Security

AI systems should protect user data and follow strong cybersecurity measures to prevent data breaches.

- AI should comply with data protection laws like GDPR (General Data Protection Regulation).
- User data should not be sold or misused without consent.

3. Transparency and Explainability

AI should be understandable to users, and its decision-making process should be clearly explained.

- Companies should provide explanations for AI-generated decisions.
- AI systems should disclose when they are being used in decision-making.

4. Accountability and Governance

There should be legal and ethical frameworks to ensure AI is used responsibly.

- Governments and organizations should set AI regulations.
- AI companies should take responsibility for AI failures.

5. Human Oversight and Control

AI should always be monitored by human supervisors to prevent unintended consequences.

- AI should not replace human judgment in critical areas like healthcare, law enforcement, and finance.
- Humans should have the ability to override AI decisions when necessary.

Challenges in Implementing Responsible AI

Despite the benefits, implementing responsible AI is difficult due to several challenges:

1. Lack of Standardized Regulations

- AI laws and regulations differ across countries.
- There is no universal standard for AI ethics.

2. Difficulty in Identifying Bias

- AI bias is often hidden in complex models, making it hard to detect.
- AI requires continuous auditing and monitoring to identify and reduce bias.

3. Trade-off Between Accuracy and Explainability

- Some highly accurate AI models are hard to interpret, making it difficult to explain their decisions.
- Developers must balance performance with transparency.

4. Ethical Dilemmas in AI Decisions

- AI must sometimes make difficult moral choices, such as self-driving car accident scenarios where it must decide who to protect.

Strategies for Ensuring Responsible AI Usage

1. Implement Strong AI Regulations and Policies

- Governments and companies should develop ethical guidelines for AI.
- AI developers should follow legal requirements related to data privacy and bias prevention.

2. Conduct Regular AI Audits

- AI models should be routinely tested for bias, discrimination, and errors.
- AI systems should undergo external reviews for fairness.

3. Provide User Education on AI Usage

- Users should be educated on how AI works, its limitations, and potential risks.
- AI-generated content should be labeled to ensure users can distinguish human vs. AI-generated information.

4. Encourage Ethical AI Design and Development

- Developers should prioritize fair, secure, and transparent AI models.
- AI teams should include diverse perspectives to reduce bias in development.

Conclusion

Responsible AI usage is crucial for ensuring AI systems benefit humanity while minimizing harm. By prioritizing fairness, transparency, privacy, accountability, and human oversight, AI can be developed and used ethically.

As AI continues to evolve, companies, governments, and individuals must work together to create AI policies that promote ethical, unbiased, and responsible AI applications.

CHAPTER 10

TOOLS AND RESOURCES FOR PROMPT ENGINEERING

Prompt engineering is an essential skill for effectively interacting with artificial intelligence models, particularly large language models. To master this skill, individuals need access to a variety of tools, platforms, research materials, and practice exercises. These resources help improve prompt design, optimize responses, and stay updated with advancements in AI.

This section explores online platforms for prompt engineering, community-driven forums, research papers, and hands-on practice exercises that can help individuals and organizations enhance their skills.

Online Prompt Engineering Platforms

Online prompt engineering platforms provide users with interactive environments to design, test, and optimize prompts for artificial intelligence models, particularly large language models such as GPT-4, Claude, PaLM, and open-source alternatives. These platforms allow users to experiment with different prompt structures, parameters, and input techniques to enhance the accuracy, creativity, and reliability of AI-generated responses.

Importance of Online Prompt Engineering Platforms

1. Real-Time Experimentation – Users can instantly test how different prompts affect AI-generated outputs.
2. Optimization and Refinement – Users can tweak prompts, adjust model settings, and observe changes in responses.
3. Accessibility – Many platforms do not require coding knowledge, allowing non-programmers to experiment with AI.
4. Business and Research Applications – These platforms help businesses automate tasks and assist researchers in natural language processing studies.
5. Customization and Integration – Some platforms allow users to customize AI behavior and integrate it into applications via APIs.

AI Model Playground and Experimentation Platforms

These platforms allow users to interact with AI models in a controlled environment to fine-tune their prompts and experiment with prompt variations, parameter adjustments, and different response formats.

A. OpenAI Playground

The OpenAI Playground is a web-based interface where users can interact with GPT models, such as GPT-4, and fine-tune their prompts in a sandbox environment.

Key Features:

- Users can input text-based prompts and receive AI-generated responses.
- Allows adjustments in model parameters, such as temperature, max tokens, and top-p, to control response creativity and length.
- Provides a history log to track prompt iterations and improvements.
- Enables different interaction styles, including instruction-based, conversational, and role-based prompts.
- Supports multi-turn conversations, helping users refine chatbot-like interactions.

Use Cases:

- Experimenting with content generation, summarization, and translation tasks.
- Testing different writing styles, tones, and formats.
- Creating chatbots, automated email responses, and interactive AI assistants.

B. Hugging Face Spaces

Hugging Face is an open-source AI community and model hub that provides access to multiple large language models (LLMs), including GPT-based models, BERT, and T5.

Key Features:

- Users can interact with pre-trained AI models for text generation, summarization, classification, and translation.
- Provides access to fine-tuning tools, allowing users to train models for specific tasks.
- Supports model comparison, enabling users to evaluate different AI models' performance on the same prompts.
- Includes a discussion forum, where AI practitioners share insights on prompt optimization.

Use Cases:

- Developers and researchers use Hugging Face to compare model performance on different natural language processing tasks.
- Businesses and AI startups integrate AI-powered solutions into applications using Hugging Face's API.

C. Google AI's Vertex AI and Bard

Google offers AI interaction tools through Vertex AI for enterprise applications and Bard, a conversational AI similar to ChatGPT.

Key Features of Vertex AI:

- Users can experiment with Google's language models, such as PaLM and Gemini.
- Provides automated machine learning (AutoML) tools to improve AI response quality.
- Supports prompt testing, fine-tuning, and integration into enterprise applications.

Key Features of Bard:

- A conversational AI chatbot that allows users to interact with Google's latest language models.
- Can generate long-form responses, creative writing pieces, and factual answers.
- Provides access to live web search results, improving the accuracy of real-time responses.

Use Cases:

- Businesses use Vertex AI for customer support automation, report generation, and AI-driven insights.
- Researchers experiment with Bard's response quality and compare it with other AI models.

No-Code and Low-Code AI Tools

No-code and low-code AI tools enable individuals and businesses to create AI-powered applications, chatbots, and automation workflows without requiring extensive programming knowledge. These tools use visual interfaces, drag-and-drop elements, and pre-built AI models to help users integrate artificial intelligence into their workflows quickly and efficiently.

Importance of No-Code and Low-Code AI Tools

1. Accessibility – Allows individuals with little to no coding experience to build AI-powered solutions.

2. Speed and Efficiency – Reduces the time required to develop and deploy AI applications.
3. Cost-Effectiveness – Eliminates the need for hiring dedicated AI developers or data scientists.
4. Integration Capabilities – Enables seamless connection with various platforms, including CRM systems, email marketing tools, chatbots, and customer service applications.
5. Customization – Allows users to tailor AI-driven solutions to their specific business needs.

1. Understanding No-Code vs. Low-Code AI Tools

A. No-Code AI Tools

- Designed for users without programming knowledge.
- Provides a graphical user interface (GUI) with drag-and-drop elements.
- Allows users to create AI workflows, chatbots, and automation tools using pre-built models.
- Examples include Chatbot builders, AI-driven workflow automation tools, and content generation platforms.

B. Low-Code AI Tools

- Requires minimal programming knowledge, usually involving simple scripting or logic-based workflows.
- Allows for customization and flexibility, making them suitable for developers and business users.
- Enables users to integrate AI models with external APIs, databases, and custom business logic.
- Examples include AI-powered application development platforms, machine learning model deployment tools, and API-based automation services.

2. Key No-Code AI Platforms

A. Chatbot Builders

No-code chatbot platforms allow businesses to create AI-powered chatbots for customer service, sales, and engagement.

ChatGPT-Based No-Code Platforms

- ChatGPT API – Can be used with no-code tools to create automated chat assistants.
- ManyChat – Helps businesses create AI-powered Facebook Messenger, Instagram, and WhatsApp chatbots.

- Tidio – Provides AI chatbots for e-commerce websites and customer support.
- Landbot – Uses a drag-and-drop chatbot builder to create interactive AI-powered conversations.

Key Features of No-Code Chatbot Tools

- Pre-trained AI models that understand and respond to customer queries.
- Customizable conversation flows that allow businesses to guide customer interactions.
- Integration with CRM and e-commerce platforms such as Shopify, HubSpot, and Salesforce.
- Multi-platform deployment, allowing chatbots to operate on websites, social media, and messaging apps.

Use Cases

- Customer Support Automation – AI chatbots can handle FAQs, process orders, and troubleshoot issues.
- Lead Generation – Chatbots can collect user data and qualify leads for sales teams.
- E-commerce Support – AI assistants can recommend products, track orders, and handle payments.

B. No-Code AI Workflow Automation Tools

These tools allow businesses to automate repetitive tasks using AI, integrating different applications without manual intervention.

Popular No-Code AI Automation Tools

- Zapier – Connects AI tools with business apps like Google Sheets, Slack, and Gmail to automate workflows.
- Make (formerly Integromat) – Automates complex workflows using a visual editor.
- Microsoft Power Automate – Enables AI-driven automation within the Microsoft ecosystem (Excel, Outlook, Teams, etc.).

Key Features

- Drag-and-drop workflow builders that automate tasks without requiring coding.
- Pre-built AI integrations that allow users to process emails, analyze data, and generate reports.
- Conditional logic and triggers, enabling dynamic AI-driven decisions.

Use Cases

- Automating Email Responses – AI can process emails and generate personalized replies.
- Data Entry Automation – AI extracts data from documents and fills spreadsheets or databases.
- AI-Powered Lead Scoring – AI analyzes customer interactions and assigns scores to potential leads.

C. No-Code AI Content Generation Tools

These platforms use AI to assist with content writing, marketing copy, and creative writing.

Popular No-Code AI Content Generation Tools

- Jasper AI – Generates marketing copy, blog posts, and social media content.
- Copy.ai – Helps businesses create AI-powered ad copy, email subject lines, and landing page text.
- Notion AI – Enhances productivity by summarizing, rewriting, and generating text content.

Key Features

- AI-powered text generation and rewriting.
- Tone and style customization for different audiences.
- SEO optimization tools to improve content visibility.

Use Cases

- Marketing Teams – Automating content creation for social media posts, blog articles, and advertisements.
- Entrepreneurs – Creating website copy and email marketing campaigns.
- Writers – Enhancing productivity by automating brainstorming and drafting.

3. Key Low-Code AI Platforms

A. Low-Code AI App Development Tools

These platforms allow businesses to develop AI-powered applications with minimal coding.

Popular Low-Code AI Development Tools

- Bubble – Allows users to build web applications with integrated AI functionalities.
- Adalo – Enables businesses to create mobile apps with AI-based automation.

- AppGyver – Provides a visual development environment for AI-powered business applications.

Key Features

- Drag-and-drop interface for building app functionalities.
- AI-powered automation that enhances user experience.
- Cloud-based deployment, allowing businesses to launch AI-powered apps quickly.

Use Cases

- AI-powered Customer Portals – Businesses can create apps that provide automated customer support and order tracking.
- E-learning Platforms – AI-driven learning tools with personalized recommendations for students.
- AI in Healthcare – Apps that use AI for patient data analysis and health monitoring.

B. AI Model Deployment with Low-Code Tools

These platforms allow users to train, fine-tune, and deploy AI models with minimal programming knowledge.

Popular AI Model Deployment Platforms

- Google AutoML – Enables businesses to create custom AI models for image recognition, text classification, and predictions.
- DataRobot – Provides automated machine learning tools to analyze data and generate AI-driven insights.
- Runway ML – Allows users to create AI-powered visual effects and art generation models.

Key Features

- Pre-built AI models that can be trained on user data.
- Cloud-based AI deployment for accessibility and scalability.
- Minimal coding required, enabling business users to leverage AI.

Use Cases

- Retail Businesses – AI-driven demand forecasting and inventory optimization.
- Marketing Analytics – AI-powered tools that predict customer behavior and market trends.
- Healthcare AI – AI models that analyze medical data for disease prediction and diagnosis.

Conclusion

No-code and low-code AI tools democratize AI development, making it accessible to businesses, marketers, educators, and entrepreneurs without requiring deep programming skills.

By using these platforms, individuals and companies can automate workflows, improve customer interactions, generate high-quality content, and build AI-powered applications quickly and efficiently.

API-Based Prompt Engineering Platforms

These platforms provide AI access through APIs, allowing developers to integrate AI models into their applications. They enable dynamic prompt generation and advanced AI-powered automation.

A. OpenAI API

- Allows developers to integrate GPT models into applications, chatbots, and business tools.
- Users can dynamically modify prompts and tailor responses based on real-time user interactions.

B. Cohere API

- Provides language model APIs focused on text generation, classification, and summarization.
- Used by businesses to build AI-driven search engines, chatbots, and content recommendation systems.

C. AI21 Studio

- Focuses on business-grade AI solutions, offering API access to advanced language models.
- Used for automating content generation, research assistance, and data-driven insights.

Conclusion

Online prompt engineering platforms play a crucial role in helping users design, test, and optimize AI-generated responses. Whether through playground tools, no-code solutions, or API integrations, these platforms provide valuable environments for experimentation, refinement, and real-world AI deployment.

By leveraging these tools, businesses, researchers, and developers can improve AI response accuracy, automate workflows, and create intelligent AI-driven applications.

Community Forums and Research Papers

The field of prompt engineering is constantly evolving. To stay updated with the latest advancements, it is essential to engage with AI research papers, community forums, and expert discussions.

A. AI and Prompt Engineering Community Forums

- **OpenAI Community Forum**
    - » A dedicated space where developers and AI enthusiasts discuss best practices for prompt engineering.
    - » Users can share successful prompts, troubleshooting techniques, and new discoveries.
- **Hugging Face Discussion Forum**
    - » Covers fine-tuning models, prompt optimization, and API usage.
    - » Users can ask questions and get feedback from researchers and AI practitioners.
- **Reddit AI Communities**
    - » Subreddits such as r/MachineLearning, r/artificial, and r/PromptEngineering provide discussions on AI advancements.
    - » Users can explore case studies, prompt crafting techniques, and real-world applications.
- **Stack Overflow and AI Stack Exchange**
    - » Developers can ask AI-related technical questions and find solutions to common problems in prompt engineering.
    - » Helpful for learning about prompt formatting, API integration, and AI-generated errors.

B. AI Research Papers and Publications

Academic research provides deep insights into prompt engineering techniques, AI behavior, and best practices. Some of the best resources include:

- **arXiv.org – AI and NLP Papers**
    - » A collection of research papers on natural language processing, AI prompt techniques, and deep learning.
    - » Users can find studies on Chain-of-Thought prompting, Self-Consistency methods, and reinforcement learning in AI.
- **Google Scholar**
    - » A search engine for academic papers related to prompt optimization, ethical AI, and model interpretability.
- **OpenAI Research Blog**
    - » Publishes insights from OpenAI on GPT models, AI alignment, and prompt-based interactions.
- **Anthropic AI Research**
    - » Focuses on safety and ethical considerations in prompt-based AI usage.

Hands-on Practice and Challenge Exercises in Prompt Engineering

Introduction to Hands-on Practice in Prompt Engineering

Prompt engineering is a skill that requires practical experimentation, iteration, and problem-solving. While theoretical knowledge provides a foundation, real-world application and structured exercises help develop expertise in crafting effective prompts for different AI models and tasks. Hands-on practice involves experimenting with different prompt structures, testing outputs, and refining prompts based on response quality and accuracy.

This guide provides detailed exercises, challenges, and strategies for improving your prompt engineering skills.

1. Why Hands-on Practice is Essential for Prompt Engineering?

- Understanding AI Behavior – Direct interaction with AI models allows you to observe how they respond to different prompts.
- Developing Effective Prompting Techniques – Practical exercises help you identify patterns in AI-generated outputs and refine prompt structures accordingly.
- Improving Response Quality – By experimenting with different prompt formats, you learn how to control and optimize AI responses.
- Adapting to Different AI Models – Each language model (GPT-4, Claude, Gemini, etc.) has unique behaviors and strengths, requiring tailored prompting approaches.
- Building Problem-Solving Skills – Challenge exercises enhance your ability to debug, refine, and optimize prompts to achieve specific outcomes.

2. Getting Started with Hands-on Prompt Engineering Practice

A. Selecting an AI Model or Platform

Before starting, choose an AI model or API to work with. Some popular options include:

- OpenAI's ChatGPT (GPT-4, GPT-3.5)
- Anthropic's Claude API
- Google Gemini AI (PaLM API)
- Hugging Face Models
- Cohere API
- Local AI models like LLaMA and Mistral

B. Setting Up a Practice Environment

- **Use AI Web Interfaces**: OpenAI's ChatGPT, Claude AI, and Google Bard allow users to test prompts in a browser.

- **Work with APIs**: Developers can use API platforms (OpenAI API, Hugging Face, Google Vertex AI) to send prompts programmatically.
- **Use No-Code AI Tools**: Tools like FlowGPT, PromptPerfect, and AI Dungeon allow experimentation without coding.

C. Basic Prompt Engineering Techniques to Master

Start with **fundamental prompt engineering techniques** before moving to advanced strategies:

1. **Direct Instruction Prompting** – Clearly specify what you want.
2. **Role-Based Prompting** – Assign the AI a role (e.g., *You are a financial analyst*).
3. **Step-by-Step Prompting** – Guide AI through structured tasks.
4. **Formatting Prompts Properly** – Use bullet points, numbered lists, and clear sectioning for better responses.
5. **Iterative Refinement** – Modify prompts based on previous AI responses to enhance clarity and accuracy.

3. Challenge Exercises for Improving Prompt Engineering Skills

Beginner-Level Exercises

Exercise 1: Basic Instruction-Based Prompting

Objective: Learn how AI interprets and follows direct instructions. **Task:** Experiment with different ways of instructing AI to summarize a passage. **Example Prompts:**

1. *Summarize this article in three sentences.*
2. *Provide a detailed summary with key takeaways in bullet points.*
3. *Explain this text to a high school student in simple language.*
4. *Rewrite this text in a formal tone for a business report.*

Goal: Observe how changing the wording affects AI responses.

Exercise 2: Role-Based Prompting

Objective: Learn how assigning a role changes AI behavior. **Task:** Ask the AI to answer the same question from different perspectives. **Example Prompts:**

1. *You are a medical doctor. Explain the symptoms of diabetes to a patient.*
2. *You are a historian. Provide a detailed explanation of World War II's impact on Europe.*
3. *You are a software engineer. Explain recursion in programming using a simple analogy.*

Goal: Analyze how AI adjusts its response based on the assigned role.

Exercise 3: Formatting and Structuring Prompts

Objective: Improve response clarity by using structured prompts. **Task:** Compare AI responses to prompts with and without formatting. **Example Prompts:**

1. *List five benefits of regular exercise.*
2. *Provide a table comparing Python and Java in terms of speed, ease of use, and applications.*
3. *Write an email template for requesting a meeting with a client.*

Goal: Understand how structured prompts improve output readability.

Intermediate-Level Exercises

Exercise 4: Few-Shot Prompting

Objective: Guide AI by providing examples within the prompt. **Task:** Compare AI responses with and without example inputs. **Example Prompts:**

1. **Without examples:** *Convert this sentence into active voice: "The report was written by John."*
2. **With examples:** *Example 1: "The cake was baked by Mary." → "Mary baked the cake." Example 2: "The project was completed by Sarah." → "Sarah completed the project." Now convert: "The report was written by John."*

Goal: Observe how AI learns from examples and applies patterns.

Exercise 5: Chain-of-Thought (CoT) Prompting

Objective: Improve logical reasoning by making AI explain its thought process. **Task:** Ask the AI to solve a math or logic problem step by step. **Example Prompt:** *"If a train travels 120 miles in 2 hours, what is its average speed? Explain your reasoning step by step."*

Goal: Understand how breaking down complex problems leads to more accurate AI outputs.

Advanced-Level Exercises

Exercise 6: Zero-Shot vs. Few-Shot Learning

Objective: Compare AI performance with and without training examples. **Task:** Use the AI to classify customer feedback as positive, negative, or neutral. **Example Prompts:**

1. **Zero-Shot Prompt:** *Classify this customer review as positive, negative, or neutral: "The product is okay but arrived late."*
2. **Few-Shot Prompt with Examples:**

- **Positive:** "The service was excellent, and the product is great!"
- **Negative:** "The product broke after one use. Very disappointed."

- **Neutral:** "The product is okay but arrived late."

Goal: Learn when few-shot learning improves response accuracy.

Exercise 7: Multi-Turn Prompt Optimization

Objective: Improve chatbot or virtual assistant responses. **Task:** Engage in a multi-turn conversation and refine the AI's responses. **Example Scenario:**

1. *User:* "Tell me about the weather."
2. *AI Response:* "Can you specify the location?"
3. *User:* "New York."
4. *AI Response:* "The weather in New York is currently 75 F with partly cloudy skies."

Challenge:

- Modify the AI's responses to improve user experience.
- Ensure AI asks clarifying questions when needed.

4. Best Practices for Hands-on Prompt Engineering

- Test Prompts with Different Models – Different AI models respond differently to the same prompts. Compare outputs.
- Use System Messages (for APIs) – Some AI platforms allow pre-defining AI behavior with system prompts.
- Iterate and Improve – Modify, test, and refine prompts for better accuracy and relevance.
- Monitor Bias and Ethical Concerns – Ensure fair, neutral, and responsible AI interactions.
- Experiment with API Parameters – Adjust temperature, token limits, and response frequency for optimized results.

Conclusion

Hands-on practice and challenge exercises are crucial for mastering prompt engineering. By working on real-world problems, testing different strategies, and refining AI responses, you can improve accuracy, efficiency, and creativity in AI interactions. Whether you are a beginner or an expert, continuous experimentation with new prompting techniques will help you stay ahead in the evolving field of AI.

INDEX